OSCEs For PLAB And Medical Students

Students

Dr Pip Fisher
Mick Sykes

PREFACE

Who is this book for?

International medical graduates preparing for part two of the Professional Linguistics Exam Board (PLAB) set by the General Medical Council (GMC), will find this book helpful as an introduction to consulting in the UK. Both authors are UK trained but have years of experience working with international graduates, so we are aware that the style of consulting in different cultures can be very different. Patients' expectations of the doctor can be very different. And the subtleties of language can be difficult to grasp. With this in mind we have included sections on the currently accepted UK consulting style to help you. You will find quite detailed guidance on language and phrasing, as well as Objective Structured Clinical Exam (OSCE) scenarios for you to practice.

UK medical students, nurse practitioners and physicians' assistants in training may have already been taught communication skills and have had a lot of opportunity to observe in the clinical environment. Nevertheless we believe you will find this book helpful both during your course and in your exam preparation. The OSCE scenarios should help you to hone your skills and make sure your theoretical knowledge is balanced by a practical ability to cover the key points in a consultation.

The topics included

The majority of the materials in this book have been used in our teaching. We have included topics that you are likely to meet both in exams and in the early years of practice. Some scenarios are written to remind you of specific points, such as the importance of checking patient compliance when medication is not working, or the importance of asking about over the counter medication or illicit drug use. Look for the boxes which highlight key learning points.

Discussion of different aspects of the consultation can be found in separate chapters that you can read through at your leisure. The OSCE scenarios themselves are cross referenced so that you can choose to work through them:

- by bodily system
- by presenting complaint (e.g. abdominal pain, breathlessness)
- or according to whether you wish to practice history,
  explanation or consent scenarios

In fact in real life you will often have to consider more than one bodily system and combine many different tasks in one consultation. You may take a history, examine the patient, make a tentative diagnosis and discuss your thoughts with the patient, all within ten or fifteen minutes. In most of the scenarios the patient can give a clear history or listen to your explanation calmly. Again in real life this is not always the case. In real life the patient may be in acute pain or confused, so you will need to calm the patient and manage initial symptoms at the same time as taking a history and beginning the investigations. Combining these tasks, and doing them well, is a complex, high level skill. Teaching and testing the ability to multitask in this way requires close one to one observation in practice, or on an advanced simulation ward, so is highly resource intensive. Nevertheless it should ultimately be the goal, so seek out feedback where and whenever you can. Just as you learn to walk before you can run, you need to learn the steps of consultation one at a time. Only once you have mastered each individual step, are you ready to put them together.

CONTENTS

INTRODUCTION

How to use the OSCE scenarios

We hope that you will use the OSCE scenarios to practice with your classmates, when you cannot find a real patient to talk to. It is not a substitute for interactions with real patients, nor is it an alternative to reading widely.

Remember the scenarios are not exhaustive. You may meet different situations and should use your time on the wards and the clinics to ensure you can handle as many different patient encounters as possible. The more histories you take and the more different procedures you watch the better prepared you will be.

> SUGGESTED ACTIVITY
> Work with two colleagues to practice the scenarios.
> One person takes the role of exam candidate; one the role of patient and one the role of examiner.
> You may want to prepare by reading about the topic on a website such as www.patient.co.uk but it is important that the exam candidate does not read the patient role or the mark sheet before trying the station. The examiner and the patient may read both parts if they wish.
> The person playing the patient should try to 'get into role' and give the responses that real patients might give. We have given you a few background details of different patients but you may want to think of more. Try to play a variety of people. Real patients sometimes don't understand doctors, or may feel angry or upset. They may be older or younger than you, come from a different class to you, a different region to you and a different ethnic background to you. It may be helpful (unless your acting skills are really good) to write on a piece of paper the age, sex and ethnicity of the patient you are playing, and keep that in front of you and the exam candidate during the consultation.
> You can vary the details from those written if you think it will improve the role. The patient should not volunteer information that is not asked for, but equally don't try to evade questions or hide information – the aim of the practice is to reinforce learning not to 'win'.
> Once you are prepared, agree a time limit and set a timer before you begin. The time limit can be generous at first, but make it closer to the time you will have for your exam stations as the exams get closer.
> When the candidate has tried the station, the examiner and the patient can give feedback and the candidate can read the patient role and the mark sheet to see where they could improve.
> Swap roles when you try other stations.
>
> Hint: International students will benefit from practising with native English speakers. If you only practice with other international students you will limit the improvement in your vocabulary and approach.

The marking schemes we have included are intended to give you a flavour of what examiners are looking for - a safe junior doctor who communicates well. We have not reproduced exactly the marking scheme of any one exam centre, but have created a general format that we hope will enable you to develop good habits. These should take you through your exams and onwards with confidence in real life consultations too.

Because we have found that many students struggle at first to pick out the important points our mark sheets are quite detailed. Do not get too stressed about the finer points but aim to develop a good general consulting style alongside a solid basic knowledge of the medicine involved. In real life different clinicians may ask questions in a slightly different order, or carry out a procedure using a slightly adapted technique. OSCE marking has to allow for this, so the order in which you ask the questions is not essential, but appearing practised and fluent is essential.

> SUGGESTED ACTIVITY
> Once you have practiced a few of our worked examples, use the generic mark sheet format provided to write your own OSCE stations. Practice this with your friends:
> - Choose a topic and each write a mark sheet that you think covers the important points.
> - Compare what you have written.
> - Did you manage to pick out the important facts that a patient would need to know in an explanation station? Or the key questions you need to ask to make a diagnosis?
> This is a good revision technique.

Don't forget the medicine

This book alone is not sufficient to get you through your exams. Even if you have the best communication style in the world, you will fail if you don't know enough to make a diagnosis, or to give an explanation, in commonly encountered clinical situations. Even worse, if you were to pass your exams without having a sound knowledge base, would you really want to go out on the wards to face real patients inadequately prepared? Would you want a close relative treated by a doctor who 'got lucky' on the days of the exams? So, practice OSCEs by all means, but never forget you need to read appropriate textbooks, attend tutorials and spend time on the wards and in the clinics, observing, listening, talking to patients and examining them.

Useful resources

Each institution will have its own favoured resources to which you will have (hopefully) already been referred. Be aware of the differences between different books and web resources and who they are aimed at. Consider varying your reading to include a range of sources.

- A good general undergraduate textbook should allow you to get an overview of a topic with an understanding of the underlying principles behind how a disease presents or a treatment works. Make sure you have one that you find understandable or you will not want to read it.
- Specialist texts or the thickest books may be useful for occasionally dipping into if you are undertaking a student selected project, but otherwise too much detail may be confusing and unhelpful at this level. Save these for later in your career.
- Revision lists and handbooks that are stripped bare of explanation, though useful for quick reference or for the night before the exams, will not help you to understand the key principles or to apply your knowledge in different situations, so are not sufficient in themselves.
- Websites aimed at patients often give a good general overview of a subject in relatively easy to understand language. This has two advantages. Firstly it is a good place to start your reading to lay the foundations. Secondly, someone else has already considered what is important for patients and how it can be explained.
- Patient information leaflets from outpatient clinics and GP surgeries can also be a good starter in helping you to prepare for OSCEs.

Websites we would recommend include

www.patient.co.uk
This site has a huge range of medical conditions and also on commonly prescribed drugs. It also provides links to many single focus websites (e.g. diabetes UK and the British Heart Foundation) which themselves provide very good information on particular medical problems. It is widely used by doctors in practice. It is a good place to start your reading. The patient information leaflets are arguably more detailed than is appropriate for most patients.

www.healthtalkonline.org
This site has lots of videos, mostly of patients talking about their experiences of all aspects of disease and medical care. It can be very useful to hear how patients feel. Often the things that patients feel are important about their care may not be the things you would have anticipated.

www.nhs.uk
This is a website with lots of information for the public, including written information and videos. It is well worth exploring.

www.gmc-uk.org/doctors/plab/advice_part2.asp#Content
Reading this site is essential for international graduates studying for PLAB, but it will also prove useful for UK students preparing for finals as it includes the General Medical Council's advice on communication skills.

www.gmc-uk.org/guidance/good_medical_practice.asp
The GMC also has a lot of other very important advice relating to currently accepted practice where the doctor is confronted by an ethical dilemma. For an easier read, try the interactive modules:
www.gmc-uk.org/guidance/case_studies.asp
and the learning materials on www.gmc-uk.org/guidance/9166.asp
These should be essential reading for everyone wanting to make sure that they keep within accepted practice when faced with a professional dilemma. International students and graduates may find them particularly helpful, as the manner in which such challenges can be dealt with varies from country to country.

www.nice.org.uk
This site contains links to many of the clinical guidelines that are most frequently referred to in the UK. They are not always an easy read, but nevertheless it is important that you learn to use these guidelines and pick out the key points to ensure that your management of patients follows current evidence- based recommendations. The pathways are particularly aimed at helping health professionals quickly determine the best evidenced approach to a problem.

www.prodigy.clarity.co.uk/home
Another well written site with evidence relating to investigation and management of many conditions, as well as links to useful patient information leaflets.

Other web sites which may be useful:
http://www.bhf.org.uk/heart-health/conditions.aspx
http://www.lunguk.org/you-and-your-lungs/diagnosis-and-treatment
http://www.asthma.org.uk/about-asthma/asthma-basics/
http://www.urol.info/start.htm
http://www.jointzone.org.uk (section on Approach to patient)
http://www.fpa.org.uk/helpandadvice (contraception)
http://www.rcpsych.ac.uk/mentalhealthinfoforall.aspx
http://www.skillscascade.com/models.htm (models of the consultation)

Tips to improve your communication skills

The general approach - patient centredness

Studies in the UK[1] have shown that you will get more information, higher patient satisfaction and better results from your consultations if your patients feel they have been listened to and understood; that their views have been taken into account and they have been involved in making the plan of action. This has led to a patient centred approach to the consultation. This is quite different from the approach used in the past in the UK, and from that used in other cultures where the patients' expectations may be different. International medical graduates may have to adapt their approach quite radically and this will not just be for the exams. This is for real life too! UK patients want and expect to be involved in decision making. In fact, used skilfully, this approach can improve outcomes all over the world, but it has to be managed with care.

If you are struggling to understand this approach, think about how you feel when an authority figure (maybe your tutor) wags a finger at you and tells you that you are wrong. You have no chance to explain or put your viewpoint. Do you feel resentful? Angry? Misunderstood? How likely are you to listen and change your behaviour? Perhaps you will agree to a plan, but will have no intention of carrying it through, just to get away? On the other hand, if that authority figure stops to find out your views, or why you behaved as you did, then how do you feel? Will you listen then?

So, being patient centred means

- trying to understand what is important to the patient
- understanding what they are worried about
- understanding what they are hoping for
- incorporating this understanding in the plan of management that you agree with the patient

Being patient centred does not mean shying away from giving advice. You can, and should, still point out the course of action that you think would have the best clinical outcome. But remember that for many situations, what is right for one person is not right for another. Talk about the possible consequences of one action or another (or none) and encourage the patient to think about how they might feel if they met these consequences. The idea is to enable the patient to make an informed choice.

Often, there is no need to reach agreement on a plan during one consultation. Where appropriate and safe for the patient, you could suggest that they go away and think the situation through, maybe talk it over with a friend or relative, and read some information about the problem. This

---

[1] For example, see Little P, Everitt H, Williamson I, Warner G, Moore M, Gould C, et al. Preferences of patients for patient centred approach to consultation in primary care: observational study. BMJ 2001; 322:468–472.

information could come from authoritative leaflets or web sites. Many UK patients are familiar with the internet, so give them the name of a reliable site that you know has good easily understood information. Make sure you agree a date for follow-up and give them advice on what to do if things get worse before they are due to be seen again.

Sometimes you may find that you have to support a patient to take a course of action that you think would not be your choice. This a good challenge to your professionalism and skills.

SUGGESTED ACTIVITY

Stop and think about how you will respond if an adult patient refuses investigation of a red flag symptom that may mean they have cancer.

Write down the key points you will discuss and compare it with our suggested course of action below.

Your response should include the following elements:

Check they really understand the importance of the test. Explain that you think the tests are needed because the symptoms could possibly be a sign of cancer, and if a cancer is detected early treatment may be possible that might not be available if the problem is left undiagnosed.

Try to find out why they do not want the test. Could helping them with transport, or changing the date, or the venue, make a difference?

If the patient still cannot be persuaded, make clear that you will continue to support them whatever they choose, and that the door to investigation is never closed if they change their mind.

Finally, document your discussion in the records.

Memorable consultations will come not from the patients who easily agree with your suggestions, but from those difficult times when you have had to accept your differences with your patient and supported them in doing things their way. Finally after many consultations, a little sign of building trust will appear and eventually you will have forged a relationship in which the patient will be able to ask for your advice when all advice from other sources might be rejected.

Ideas, Concerns and Expectations

A commonly used mnemonic that will help you remain patient centred is ICE. It stands for:

Ideas   This is what the patient thinks is the problem
                     'So what do you think is going on?'
Concerns   This is what the patient is worried about
                     'You look awfully worried. What is it that is   worrying you so much?'
Expectations   This is what the patient hopes will happen as a
                     result of the consultation
                     'What were you hoping I/ the doctor would do for you today?'

Students sometimes struggle to ask about ideas concerns and expectations in a way that sounds natural. There is a danger that the questions will sound flippant and elicit equally flippant responses, for example, the question
    'So what do you think the problem is?'
may elicit the response
    'I don't know, you are the doctor!'
Asking
    'What do you expect me to do about it?'
is particularly risky, if you are not very careful about your tone of voice and manner of delivery. It can be taken as quite rude in Britain; a phrase used by truculent teenagers responding sulkily to being scolded by their parents. The term *hope* is much less likely to cause problems.

You will need to be very careful in wording your questions and in slipping them into the conversation at the right moment. Sometimes it is helpful to ask these questions when you are examining the patient. The examination helps to distract the patient and the questions sound more relaxed. An alternative approach is to consider whether your patient might have spoken to someone else or looked up the symptoms on the internet before coming to the doctor, and asking about that.

'Did you look this up on the internet/ talk to your husband/friends about this?'
Remember also that you can gather a lot of information by just listening, allowing the patient to open up, using silence and reflecting the patient's words. You do not always have to ask the ICE questions directly.

Effect on daily life

You will notice that in our mark sheets for history taking, we include both ideas concerns and expectations and effect on daily life. We have put this in because again what is important to the patient may not be what is important to you. Here are a couple of examples but you could find many more:

A patient consults with symptoms of carpal tunnel syndrome. You ask how the symptoms are affecting the patient's daily life and learn that he is a concert violinist with a big competition coming up. You can understand how worried the patient may be and express this. You discuss the treatment options and find he is keen to try the options that are likely to get him back to top form in time for the concert. He chooses a steroid injection into the wrist. Your patient leaves satisfied that you have understood his point of view.

A different patient comes in with symptoms of carpal tunnel syndrome. She is pregnant and has gained weight during the pregnancy. She is afraid that the pain will stop her being able to cope with the baby. You can explain that her symptoms are likely to subside once she has the baby and loses the extra weight, so she is happy to be managed conservatively with wrist splinting for now.

Taking a history

So far we have focused on gaining the patient's perspective. We have put a lot of emphasis on this, as it is often neglected in the rush to gather the information required to make the diagnosis. It might not have been considered in your training if you trained years ago, or in a different country. However clearly there is more to consulting than just this aspect.

There are lots of models of the consultation written by wise and experienced authors and we do not wish to reproduce them all here. For a quick overview of the different models you might like to look at www.skillscascade.com/models.htm

The Calgary Cambridge model by Silverman, Kurtz and Draper[2] and has become very popular and their book includes a useful pocket sized fold up sheet that you may wish to use as a quick reminder during your practice sessions.

Remember models are designed to help you organise your thoughts and check you have covered everything. There is no one right or wrong model to use. We have designed our history mark sheet (see p353) and the more exhaustive crib sheets (see p356) to include the aspects that we feel are important

If you wish to take notes whilst taking a history, you might find it useful to photocopy our crib sheet and write on that, but a more succinct note taking method is to divide your page into 4 sections:

- Presenting complaint (including relevant associated symptoms)
- Patients ICE and effects on daily life
- Past medical history, drugs, allergies
- Social history

Let the patient do the work

When we are nervous, many of us talk too much. You will be nervous in your exams and sometimes when you are talking to patients, so you will have to learn to fight this tendency and to allow the patients to do the talking.

Beginning your history taking with an open question allows the patient to start talking and avoids assumptions. For example,

'How can I help you today?'

or *'What is the problem today?'*

is better than *'How long have you had your chest pain?'*, even if you have already read on the station description that the patient has chest pain.

Following up with

'Can you tell me a bit more about that?'

---

[2] Silverman JD, Kurtz SM & Draper J (1998) Skills for Communicating with Patients. Radcliffe Medical Press.

invites more description of the issues involved and saves you a lot of work. Open questions (that cannot be answered in one or two words) allow the patient to discuss their fears as well as their medical symptoms.

Start with open questions ...

... move on to more closed questions to clarify details

When discussing test results, asking
> *'Do you know what we were looking for?'*

will allow you to find out the patient's understanding and concerns, before you have even started to give any information.
Equally
> *'Do you know what this procedure involves?'*

may save you lots of time if the person you are about to consent for endoscopy turns out to be an endoscopist.

Use silence to encourage the patient to expand on what they have already said. British people are very polite and often find silence in a conversation embarrassing, so will talk to fill the gap. This is particularly the case if talking about something emotionally sensitive. Let the patient fill the gap, not you.
An alternative to straight silence is to repeat (or reflect) the patient's last phrase back to them, then pause. You might find this less uncomfortable than simply saying nothing.
So talking about what a patient thinks might be causing the symptoms, your conversation might go as follows:
> *'So what do you think is going on?'*
> *'Well, you know I've been a smoker all my life so...'*

You now have a few choices.

- You could go on to ask how many the patient smokes, but this gives him no chance to talk about his fears and effectively ignores the cue he is giving you about his fears. In an exam you would lose marks and in real life you would lose the patient's trust.
- You could just pause and wait. In a few seconds the patient will start talking and may well mention directly a fear of lung cancer.

- You could repeat '*...so...?*' which similarly is likely to lead the patient to expand on his fears.

When you are successfully using silence and reflecting the patient's words, you will know that you are achieving a high level of communication skills.

Moving on to obtain more information

Here are our hints on completing your history taking.

History of presenting complaint

The mnemonic SOCRATES can provide a useful reminder when you are taking a history of the presenting complaint. Although it is used most often in relation to pain, you can adapt it to other symptoms

S – site

O – onset (gradual or sudden, triggered or out of the blue)

C – character

R – radiation

A – associated symptoms

T – timing/duration

E – exacerbating or relieving factors

S – severity

Associated symptoms

When you are asking about associated symptoms, an open question such as

'Did you feel unwell in any other way at the time?'

is useful as a starting point. However you are likely to want to know specifically about the other symptoms that often go along with problems in that bodily system and may need to ask a few closed (yes/ no) questions to obtain this information. For example, if you are thinking of a cardiac problem, you will want to ask about all the symptoms that come listed under cardiovascular history in your crib sheet: chest pain, palpitations, shortness of breath, ankle swelling and claudication, as well as possible about nausea and sweating (related to chest pain). And remember chest pain and breathlessness could be cardiac or respiratory so you may want to ask questions relating to both the cardiac and respiratory systems.

Information obtained by asking direct, closed questions may carry slightly less weight than a description of symptoms the patient volunteers in response to an open question. This is because patients tend to be very polite and will sometimes agree just to please the doctor.

As you get more proficient, your history taking may stop following the sequence of the crib sheets. You should start to think of differential diagnoses quite early in the history, then use your questions to help you to decide which of those possibilities is more likely. For example, a patient presenting with bloody diarrhoea may have infectious dysentery, inflammatory bowel disease, colorectal cancer or piles. So quite early on in the history you should be asking about infectious contacts and recent travel, any family history of bowel disease, any joint or eye problems (which may occur in inflammatory bowel disease) and any weight loss (which could occur with each of the more serious conditions). This saves time and shows the examiner that you are thinking of the differential diagnoses.

Past medical history

Useful phrases here include

> *'Have you any other medical problems?'*
> *'Have you had any operations or ever been in hospital at all?'*

But it is common for patients to forget to mention illnesses which you might consider quite serious, so it is worth outlining the medical problems that you want to know about specifically – diabetes, epilepsy, asthma, heart problems, blood pressure…

Drug history

British people tend to think of the word *'drugs'* as meaning drugs of abuse such as heroin, cannabis, cocaine, so you might want to ask

> *'Are you taking any tablets or medicines?'*

BUT don't forget that many people also take non-prescribed medication, such as herbal remedies or vitamin supplements. AND many people in the UK take drugs of abuse, so you will need to ask about them.

> *'Are you taking anything that you have not been prescribed?'*

should tackle both of these. However don't be afraid to ask

> *'Do you take any recreational drugs?'*
> *'Do you take any street drugs?'*
> *'Do you do drugs?'*

Another approach might be to tack the question about illicit drug use on after questions about smoking.

> *'Do you smoke anything else?'* will pick up on cannabis use

Or you could ask try after asking about alcohol.

> *'Do you use any drugs?'*

Smoking and alcohol

Again don't shy away from asking about these. When asking about alcohol, try to pin the patient down to what they drink, how many drinks and how many days in the week, rather than accepting answers such as *'not much'* or *'socially'*. Don't ask patients how many *units* of alcohol they drink. This is for you to calculate, not a question for the patient. And if you are not certain how to calculate it then just write down the type of drink taken, number and frequency. You can make the calculation later.

Social history

International students in particular often struggle to ask about home circumstances – UK patients may be married, living together, heterosexual, homosexual, single, divorced, widowed, with or without children or any combination of these. An easy approach to asking about this is simply;

'Who is at home with you?'

That avoids any unintentional, implied judgement.

If your patient is of working age, ask

'Do you go out to work?'

'What is your job?' 'What does that involve?'

Remember issues such as travel history or pets and unusual hobbies, especially if you are thinking of infectious diseases or other illnesses where these risk factors may be as relevant as the patient's job.

Family history

Again you may want to specify which illnesses you are thinking of, according to the complaint the patient presents with.

Systematic enquiry

Students often struggle with the systematic enquiry at the beginning and worry that they will not have time to cover everything in the exams. Again it is important to remember to give priority to the questions RELEVANT to the presenting complaint. Only if you have time in an exam will you be expected to cover the rest, and in real life doctors do take short cuts, so you can do that too if time is pressing:

CVS	*'Have you had any chest pain or problems with your heart?'*
RS	*'Do you have any problems with your chest or your breathing?'*
GI	*'Are you eating OK?'*
	'Any tummy pain, or indigestion?'
	'How are your bowels?'
	'Is your weight going up, going down or just the same?'
Urology	*'Do you have any problems with your waterworks?'*
Gynae	*'Are your periods regular? How often do you have one?'*
	'Can I ask what you are using for contraception?'
	'Any chance you might be pregnant?'
CNS	*'Do you have any problems with your eye sight?'*
	'Any problems with your hearing?'
	'Any fits or faints or funny turns?'
MSK	*'Do you have any problems with your joints?'*

If asking the systematic enquiry feels like a long list of closed questions (which in truth it is) then explain to the patient:

'Now I have a list of questions I have to ask to make sure I get all the information I need. Is that OK?'

Make sure you really understand what the patient means.

Many terms are used by different people to mean different things, so it is essential that you ask a patient to clarify exactly what they mean. We have listed a few of the terms that commonly cause confusion, but you might hear others:

When a patient says...	You might like to ask ...
'I had a fit/ a blackout'	*"What would I have seen if I had been watching?'* *'Did you fall to the floor?'* *'Could you hear people around you or respond to them?'* *'Did you bite your tongue?'* *'Did you wet yourself?'*
'I've got diarrhoea'	*'How many times a day are you going?'* *'What does the poo look like?'* *'Is there any blood or slime/mucus in it?'* *'Can you see food in it?'*
'I'm going to the toilet all the time'	*'Is that for a wee or a poo?'*
'There is no chance I am pregnant'	*'Is that because you are not having sex, because you use contraception absolutely 100% of the time or because you don't want to be?'*
'I don't drink much / only drink socially'	*'How many times in a week do you drink?'* *'What do you drink?'* *'How many do you drink?'* *'Anything else?'*
'I've had palpitations'	*'Can you tap out the rhythm?'* *'Was it fast and regular, like a missed beat, or all higgledy-piggledy?'*
'My periods are irregular'	*'How often do you have a period?'* *'What is the shortest time from the start of one to the start of the next?' ' And the longest time?'* *'When did your last period begin?'* *'How many days did you bleed for?'*
'I feel dizzy'	*'Are you feeling light-headed or is the world spinning?'* *'Is it all the time or just if you move your head?'*

Summarising for clarification

Summarising what the patient has said is a valuable technique you can use to check that you have understood correctly. It also gives you a moment's pause to structure your thoughts and take stock of where you want the consultation to go next. Be clear to the patient why you are doing this. Use a phrase such *as*

> *'So just let me check I have understood everything you have said so far. You said... is that correct? Is there anything I have missed?'*

Listen and respond

At first you may be so focused on asking the questions you have learnt as part of your model history, that you don't really listen to the patient's answers. Most examiners are familiar with the situation where a simulated patient says

> *'I am really worried about...',*

and the student, not really listening, goes on to ask

> *'Do you smoke?'*

(or something else equally irrelevant at that point in the conversation). If you have a chance, try to observe yourself on videoed consultations, you will soon see the effect this has. Simulated patients are often quite kind and try to raise their concern again later, real patients may not be this persistent. They may well feel that they tried, but that the doctor did not feel that their worries were important. Overcoming this bad habit takes a lot of practice. You will need to be extremely familiar with the structure of a medical history, so that your mind is not always on the next question.

Remember the best consultations, like other conversations, are a little bit unpredictable. It may be that you have to follow the patient's lead. Sometimes it is more important that you gain the patient's trust than that you make the diagnosis that day. From an exam point of view, you will rarely fail an exam if you stopped to comfort a worried patient or calm an angry patient, even if this means you do not get a full history. However, you may fail the exam if you fire off lots of questions but the patient feels their worries have been ignored.

If you really need more information before you address the patient's fears, for example if the patient is worried their symptoms point to cancer but you have not yet gathered enough information to hazard a diagnosis, then say so.

> *'You are obviously worried about that. Just let me ask a few more questions then I will know more and we can talk about your fears.'*

This acknowledges the patient's worries and reassures them that you will return to the topic later.

Make sure the patient really understands you

It goes without saying that you should avoid using medical jargon when talking with patients. It is surprising how difficult this can be as you soon become used to using technical terms that only a few months ago may have sounded completely alien to you. Talking in appropriate language is a skill you need to practise. It is acceptable to use a few medical terms as long as you explain them. For example, if you are going to send a patient for a bronchoscopy, it is helpful for the patient to know the word (and, obviously, what it means).

There is no harm in telling the patient to stop you at any time if there is anything you say that they don't understand, but be aware that many people will be too embarrassed to tell you they don't understand. Watch their body language and respond if they look confused.

'You look a little confused. Have I said something you don't understand?'

It can be helpful to outline to the patient the direction that the conversation is going to take. This is termed signposting. It works a bit like chapter headings in a book, they tell you what to expect. You might say for example,

'First I will tell you what the test showed and then we can talk about the treatment that is needed'

Breaking up your explanations into small bite sized chunks is helpful. You can then check the patient understands each bit, before moving on to the next bit of information. For obvious reasons this technique is often known as chunking and checking.

Some schools teach students to ask the patient to summarise the key points back as a means of checking understanding. This technique is used because, as we said earlier, most people are far too polite to speak out when they don't understand the doctor. However it can feel a bit artificial, so if you are going to use this technique you will need to introduce it in a manner that you feel comfortable with. It often helps to be honest with the patient and admit that you want them to repeat the key points back to you so that you can be sure you have told them everything you should. Breaking bad news is a particular case, when you might want to ask who the patient has to support them and what they will tell them when they get home.

Talking about embarrassing topics

Many students find talking about subjects such as sex, drugs of abuse, male and female genitals, periods, and even death, embarrassing. This is only to be expected as most societies have taboos around discussing such topics in polite society. BUT in medicine there will be times when it is ESSENTIAL that you obtain the correct information from the patient and give the correct advice, so you will HAVE to discuss them. It is useful to remember here that the patient will be more embarrassed if you appear shocked, shy or inhibited. You are the professional in the consultation, so you have to appear totally at ease discussing even topics that you would never discuss outside the consulting room. Even if the patient tells you that they have sex swinging from the ceiling fans dressed in an elephant costume!

To avoid shocking the patient when moving on to potentially embarrassing questions, it may be useful to reassure the patient that the questions are routine in such cases. Signal to the patient that you are going to ask about something sensitive, for example,

> *'Now I need to ask some specific questions, to help me understand a little more about your problem.'*

Generally using euphemisms, such as *'down there' 'privates' 'ladies bits'*, implies that you are embarrassed and is best avoided. There is a danger that you will be meaning one thing and the patient another. Using anatomical terms may well be the safest and most acceptable option for body parts (vagina, penis, breasts), but terms such as defecation, urination or coitus should be replaced by their everyday counter parts, poo, wee, or sex for clarity.

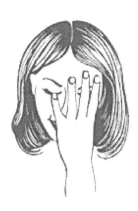

Discussing results and breaking bad news

Occasionally students appear to see explanation stations as a chance to tell the patient (or the examiner) everything they know about a topic. This is not the purpose of this station. Spewing out lots and lots of facts in the face of a bewildered patient is not what is required in exams or in real life. Your explanations must be appropriate to the patient, allow them to ask questions and to gauge what your news means for them.

Several strategies can help you (and the patient) when giving news of results, especially if the patient might consider the results to be bad news:

- Find somewhere private to give the news if you can. Consider whether the patient would like a supporter, for example a relative or nurse, with them to hear the news, if this is possible.

- Start by finding about how the patient has been since the tests were done and how they are feeling right now. If your news is going to be serious, finding out a little about their social circumstances early in the consultation will be helpful later on. Perhaps they have a partner or friend in the waiting room they would like with them during the consultation?

- Find out what they are expecting.
  'Do you know what we were testing for?'
  This will tell you how much of a shock the results are going to cause the patient. For example, if the patient think the test results are going to show an infection and they actually have cancer, you know that you will have to lead them a long way. If, on the other hand, the patient mentions cancer themselves, they will have done much of the hard work of the consultation for you already.

- Break the news in small steps, with pauses in between. We have already called this technique 'chunking and checking'. An example might be, -
  'The test showed an abnormality (pause)'
  'We do have some treatment for it, but it is serious (pause)'
  'Is that what you were expecting? (pause)'
  'The test showed that the lump you found is a tumour (pause)'
  'I am afraid the tumour is a type of cancer (stop speaking)'.
  Doing this helps the patient keep up with you and allows them to check anything they are not sure about. Use the pauses to watch them carefully to see their reactions. Respond to them during this time.

- After you have given the news, let the patient react and offer appropriate support and sympathy.

- In breaking bad news, you must be clear and accurate. If the diagnosis is cancer, the patient should know this and you should use the word cancer at an appropriate point. If the diagnosis is likely to be serious, but you are not yet 100% sure, then it is important that the patient knows both these facts.

- Give the patient a chance to ask questions and watch the body language. If they look confused or afraid, use your observation to allow them to express themselves

  'You look confused. Have I said anything that you don't understand?'

- If you give too much information at once, the patient is unlikely to remember it. Tell them where they can get more information and support (whether that is from yourself or colleagues, friends, leaflets or web sites). Knowledge of their social circumstances will help here, and you might want to discuss ways they will tell relatives and close friends.

- Make a plan with the patient for what happens next. This will help the patient considerably, especially if the news has been bad. They are likely to have questions they only think of once they leave the consultation, so will appreciate a chance to come back and ask later.

We have tried to structure our mark sheets for the explanation stations to help you develop a strategy to tackle these situations.

Safety netting

If you are making a diagnosis and advising the patient of the next steps, it is important that you think of what might happen if the patient's condition deteriorates. This is particularly important in the community, as there will be no nurses to observe the patient's condition. To ensure the patient knows what to look out for and when to get help again, you will need to agree a plan. This is termed safety netting. Safety netting means ensuring that the patient knows exactly

- what to look out for as a sign that the condition is getting worse and more medical help is needed
- how to get appropriate help if things do get worse

Good safety netting can save the patient's life and your career. For example, a GP will see many, many children who have a bit of a fever and are a bit miserable but have no more specific signs. The vast majority of these children will get better within a day or two with no medical intervention. If all these children were referred for tests the NHS would grind to a halt and lots of children would have unnecessary tests. BUT rarely a child may have a fever and be miserable because they are in the early stages of a life threatening illness such as appendicitis, or meningococcal septicaemia. The good GP will say to the parents of non-specifically febrile children

'It is probably just a viral infection and will get better in a couple of days, but if it does not, if it gets worse, or your child develops a rash or bad pain, then call the doctor again. We will be happy to check him again'.

Safety netting is particularly important during telephone consultations, as you are lacking the visual clues that might otherwise help you to a diagnosis, but you can use it during face to face consultations too, even if you are already planning follow up.

'So we will book an appointment to see you again in two weeks' time when we should have the results, BUT if it gets worse you can always come back to us before then.'

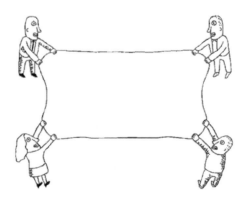

Dealing with frightened patients

Patients are very often frightened when they come to the doctor, and reassuring them is one of our most important roles. You can be reassuring without making false promises or lying about the situation.

- It is helpful to the patient to know that there is a doctor in charge who has dealt with similar cases before and is not themselves shocked or fearful, which is why looking calm is so important.
- Find out exactly what the patient is afraid of and your reassurance will have considerably more value. For example, if they know of someone who had a bad experience years ago, reassure them that things have changed since then.

When fear may be justified, for example if they are facing a serious illness, then phrases such as

> *'It is serious but you are in the right place to be treated for it now'*
> *'We are here to help you get through it'*
> *'We are going to look after you'*

can be helpful.

If treatment is palliative rather than curative, it is important to be clear about this to the patient but again to take away the fear.

> *'We don't have a cure but we do have treatment to help you'*

Many patients may be more afraid of dying in pain or breathless, or of being dependent on others, than they are of death, so again you need to find out what their concerns are and address these.

Dealing with angry patients and relatives

Inevitably during your career you will meet some angry people. Sometimes this anger may be justified when mistakes have been made, sometimes the anger will be misdirected. Remember that patients and their relatives are likely to be feeling upset, worried and afraid because of illness or the fear of illness. Hospitals can make this worse, as they can be frightening, alien environments to those who don't work in them. We all react more rapidly when we don't feel in control and are frightened of what might happen, so often this explains why people behave as they do. With this in mind, when faced with an angry patient or relative it helps to:

- Find a quiet calm place off the ward in which to deal with the situation
- Sit down if possible – this shows you are giving the complaint time and it is less provocative than standing face to face
- Always be polite, and show that you are willing to listen (however provocative the situation)
- Allow the complainant time to vent feelings
- Acknowledge the complainant's feelings
- Try to understand the reasons behind the anger

- Apologise for how upsetting the situation is
- Try to move the conversation on to how the patient can be helped now – agree a positive plan of action

Do not:

- take personal offence at anything that may be said
- make derogatory comments about colleagues
- inflame the situation by directly contradicting the patient, interrupting an outburst or behaving in a provoking or threatening way
- make promises that cannot be kept

Acknowledging and apologising for a distressing situation is important in UK culture. It does not mean loss of face, nor does it necessarily imply an admission of guilt, although you should admit if a mistake is known to have been made. Frequently the stress of a prolonged drawn out, formal complaint can be avoided if there is

- acknowledgement and apology for the situation
- action to deal with the outcome
- evidence that lessons will be learned to prevent mistakes happening again if they have occurred

Junior doctors may not be aware of the details of complaints procedures where they are working, but they MUST be aware that each organisation has a clear pathway for dealing with complaints. If you don't know about the procedure where you are working, always offer to find out and assist the complainant with the next step. Asking a senior colleague to speak to the complainant is likely to be a sensible first move.

> SUGGESTED ACTIVITY
> Why not go and speak to someone from PALS, the Patient Advocacy and Liaison Service, in the organisation you work in, and ask them about common complaints they hear and how they are dealt with.
>
> Read the GMC's advice about being open and honest with patients if things go wrong (Good Medical Practice, General Medical Council, London, point 30, available at:
> www.gmc-uk.org/guidance/good_medical_practice/relationships_with_patients_ope n_and_honest.asp).

Communicating with colleagues

When you are providing information about patients to your colleagues, you need to be very clear, so that the patient gets the care they need. It helps to have a structure to communicate this information, just as it helps to have a structure when you are taking a history or breaking bad news. A useful mnemonic that you may wish to use when presenting a case, for example on a ward round, or in summarising a case in the records, is SOAP:

Subjective	What the patient complains of
Objective	What you found (key positive and key negative findings)
Assessment	Your clinical impression/diagnosis
Plan	What you intend to do now

For more complex communication, especially when this is not face to face, the NHS is gradually adopting a structure called SBAR, to assist in communication between colleagues, for example when making a phone call or a referral. This is outlined below:[3]

S Situation:
    Give your name and the place you are calling from
    Give the patient's name and the reason for your report
B Background:
    Give the patient's reason for admission
    Give significant medical history and significant events since admission
A Assessment:
    Airway status
    Breathing (including respiratory rate and peripheral oxygen saturation)
    Circulation (including heart rate and blood pressure)
    Disability (conscious level, blood glucose and pain)
    Exposure and examination from top-to-toe
    Clinical impressions
    Actions taken (for example, oxygen and IV fluids prescribed)
R Recommendation:
    Explain what you would like the other person to do
    Agree a time frame for action(s)
    Clarify expectations

---

[3]There is a helpful resource at:

http://www.nottingham.ac.uk/nmp/sonet/rlos/patientsafety/sbar/
which provides examples of the use of SBAR

Telephone calls

Telephone stations can come up in both PLAB 2 and medical student OSCEs, and the scenarios usually involve discussing a patient with a senior colleague. In real life too, telephone calls can be tricky as there is no body language to help you, so you must ask for clarification and feedback. This is true of anyone, but if you have a strong accent, or English is a second (or third ...) language, it is vital.

SBAR is a very useful way of structuring telephone calls about patients. At the end, summarise what you think you have agreed back to the other person, to check that you have understood and have been understood. If there is any confusion over words or names, spell them out. Use a simple system such as 'a for apple, b for book,' and so on. Be especially careful with easy to confuse words such as hypo/hypertensive, can/can't. If in doubt, use 'low/high blood pressure', 'able to/not able to'.

If English is not your first language.

If you English is not your first language, you are likely to find you have to work harder than native English speaking colleagues to reach the same level of skill. This is only to be expected. You will have the advantage when patients from your first language community need help.

- Practice with native English speakers is essential, if you are to pick up the subtle phrases and tips that will help you to sound perfectly fluent. If you only practice with others who have English as a second language you will simply repeat their mistakes.

- Ask your colleagues to correct your grammar and pronunciation - they may need some encouragement as they may feel it would be impolite, but reassure them that it is in your interest to be corrected early on.

- As already mentioned, be particularly careful to find out the correct words to use around topics such as sex, urination, defecation, menstruation, death and dying. There are different words to use when talking to colleagues and when talking to adult patients and child patients. Make sure you know the words and the appropriate times to use them.

SUGGESTED ACTIVITY

If English is not your first language, get together with native speakers and play a game in which they write down lots of terms for different body parts and you have to pin them on to a diagram. How many can you get right?

Play the game a second time aiming to divide the words into groups: a group of words the doctor can use with the patient; a group that the doctor must understand, but which the doctor should probably only use if the patient uses them first; a group of words which the doctor may hear (on the street or in the pub) but should never use with a patient; and finally words adults use when talking to children.

Cardiology practice scenarios

1 History   Candidate role

You are a Foundation doctor in General Practice.

A 47 year old man has come to see you complaining of breathlessness on exertion.

Please take a history. When you have finished please:

- tell the examiner your differential diagnoses, with the most likely first,
- outline any investigations you would request.

1 History   **Patient role**

You are a 47 year old man.

You are troubled by breathlessness, which you get when you exercise. You have come to the see your GP because of it.

The breathlessness started a month ago – when you started to go jogging. You took up jogging because you finally decided to do something because you are getting a bit fat in the stomach and the furthest you had walked for some time was between the house and the car, or between the car and your office (you are an accountant and you always take the lift to your ninth floor office).

You only managed to go out for a jog twice, and you only got to the end of the road before you had to stop with breathlessness. Since then, you have tried to exercise by going for walks, but as soon as you have to go up a steep hill, the breathlessness comes on and only goes away when you rest. Sometimes you feel a tightness across the front of your chest when you get the breathlessness.

You are very worried about this - 47 years old is far too young to be unable to do exercise – maybe there is something in your lungs that is causing this? A growth maybe?

You used to be quite fit when you were younger – you played cricket in the Lancashire League for a few years in your twenties.

Background

You have not noticed any other symptoms and have had no major illnesses – minor fractures to bones in your right foot from a cricket ball years ago was the only time you have been in hospital.

You live in a bungalow with your wife.

You have one daughter who is a third year medical student in Manchester.

You have never smoked cigarettes or drunk alcohol (well, not since you were at University over twenty years ago, anyway).

Your father died of a stroke when he was 62 and one of his brothers is disabled from a stroke he had while recovering from a heart attack in hospital. Your mother is alive and well. You have one sister who has no health problems that you are aware of.

1   History   **Examiner mark sheet**

Introduces self and role and checks patient identity
Explains purpose of interview
Establishes nature of presenting complaint
- onset
- time duration
- exacerbating and relieving factors
- severity

Enquires about relevant associated symptoms:
Cardiac
- chest pain
- palpitations
- nausea
- sweating
- ankle swelling
- claudication

Respiratory
- cough
- wheeze
- sputum
- haemoptysis

Establishes effect on daily life
Establishes patient's concerns, validates and addresses these (fear of having a growth in their chest)
Drug history (including over the counter and illicit drugs)
Allergies
Establishes previous medical history
Establishes family medical history
Establishes Social history:
- occupation
- who is at home
- smoking
- alcohol

Excludes other systemic symptoms
Appropriate questioning technique (mixture of open and closed questions)
Avoids or explains jargon
Uses tools such as signposting and summarising
Systematic, logical approach
Checks if the patient has any other questions
Friendly approach, appropriate body language
Finishes consultation appropriately

1 Notes

Differential diagnoses

This patient may simply be unfit but you may want to exclude angina.

Investigations

FBC (exclude anaemia)
Fasting lipids – to help calculate cardiovascular risk
Fasting glucose
U&Es – cardiovascular disease risk often goes hand in hand with chronic kidney disease
ECG – resting ECG may be helpful in scoring the likelihood of coronary artery disease but a normal resting ECG does not exclude angina.
A trial of a GTN spray may help you to distinguish between the tightness of angina pain and simple breathlessness.
Referral for CT angiography or non-invasive functional testing (e.g. stress echocardiography or MRI) may be necessary if you feel the breathlessness described could be the tightness of angina and you score the patient as having a likelihood of coronary artery disease of 10-60%.

LEARNING POINT
NICE guidance for assessing chest pain www.guidance.nice.org.uk/CG95 provides a clear pathway to follow IF your assessment of this patient is
Breathlessness = tightness = angina pain
Scoring of the likelihood of angina takes into account, sex, age, history (typical or atypical of angina), smoking, hyperlipidaemia and diabetes.
BUT your assessment will depend very much on how the patient role is played. We are NOT stating here that this is clearly a case of angina that should be referred for investigation. The NICE guidance itself states that the scoring system is likely to overestimate the probability of coronary artery disease in a primary care population.
This case illustrates the difficulty that we face when we try to apply guidelines, which present symptoms as very clear cut, to real life in which different patients present in different ways, which do not fit easily into one diagnostic category. It is tempting to simply investigate every case in order to avoid missing a case of angina, but remember that investigations in themselves are not harmless. They cause stress, anxiety and in some instances physical harm, as well as using up finite resources. Ultimately there will always be a place for using your clinical judgement in such cases.

---

2 History   Candidate role

You are a Foundation doctor in Emergency medicine.

Your next patient is a 49 year old, who has been admitted via ambulance, complaining of chest pain.

Please take a history.

At the end of the history, summarise the key points to the examiner.

2   History   **Patient role**

You are 49 years old.

You had a pain in the centre of your chest behind your breastbone, like something was squeezing. It came on early this morning while you were in bed (about 3 am). It didn't go off with anything that you did – sitting up, lying down, walking around, or the two paracetamol you took. It was so bad it made you feel sick and sweaty. By 6 am you and your spouse were very worried and called an ambulance which brought you to the hospital. You are very relieved that the pain went after you were given a spray under your tongue and then an injection in the Emergency department. You are now on the Admissions ward.

You have never had pain like this before. If you are asked to score the pain, it was about 8 out of 10 earlier in the morning, but it is now about 1. You are very worried that you might be having a heart attack today and you might die.

Background

Your only other medical problem is mild arthritis of the hands and knees, for which you buy paracetamol from the chemist when you need it.

You smoked 10 – 15 cigarettes a day (never a full packet) from the age of 15 until two years ago, when you stopped. You drink an occasional glass or two of wine with dinner (about twice a week) - you've heard that it is good for your heart. You also like a glass or two of beer at the weekends.

You are a plumber and live with your spouse and two of your four children, all of whom are healthy. Your daughter and youngest son are in their early twenties but seem reluctant to leave home. The two eldest sons are both married.

Your two sisters both had high blood pressure and one also has diabetes (if asked, you don't think she uses injections).The other sister died of breast cancer a couple of years ago. Your mother died of a stroke at age 69, your father died of bowel cancer at age 82.

You take no medicines regularly and have no allergies that you are aware of.

2   History   **Examiner Mark sheet**

Introduces self and role and checks patient's identity

Explains purpose of interview

Establishes nature of presenting complaint

- site
- onset
- character
- radiation
- time duration
- exacerbating and relieving factors
- severity

Enquires about relevant associated symptoms:

- breathlessness
- palpitations
- nausea
- sweating
- ankle swelling
- claudication
- cough /wheeze/ sputum/ haemoptysis

Establishes effect on daily life

Establishes patient's concerns, validates and addresses these (fear of having a heart attack and dying)

Drug history (including over the counter and illicit drugs)

Allergies

Establishes previous medical history

Establishes family medical history

Establishes Social history:

- occupation
- who is at home
- smoking
- alcohol

Excludes other systemic symptoms

Appropriate questioning technique (mixture of open and closed questions)

Avoids or explains jargon

Uses tools such as signposting and summarising

Systematic, logical approach

Checks if the patient has any other questions

Friendly approach, appropriate body language

Finishes consultation appropriately

2 Notes

> LEARNING POINT
> Summary
> You should finish this station with a two to three sentence summary, such as this:
> *This is a 49 year old wo/man presenting with a three hour history of squeezing chest pain, that came on in bed in the small hours and was only relieved when s/he received a spray and an injection in hospital. This was associated with nausea and sweatiness and the fear of dying. It is the first episode of such pain, in a patient who has no past medical history of note but has a smoking history of 15 pack years.'*

> LEARNING POINT
> How did you handle the patient's question about dying? As a doctor you must be able to reassure and calm the patient, without being dishonest. You might find phrases such as
> 'You are in the right place to get treatment now'
> 'We are going to do all we can to look after you.'
> are useful. Also, remember that heart attacks had a much higher mortality in the past but current treatments mean far fewer people die. Tell this to your patient.

3 Explanation   Candidate role

You are a Foundation doctor on the cardiology ward. Your next patient is 56 years old. S/he is due to be discharged from the ward following treatment for an ST elevation MI. The patient has seen the cardiac rehabilitation nurse already, but the nurse told you she was called away before she had time to talk about the discharge medications.

Please talk to the patient, to make sure that s/he knows enough about the medicines on the discharge sheet.

Medications:

Aspirin 75mg od
Bisoprolol 5mg od
Clopidogrel 75mg od
GTN spray prn
Ramipril 10mg od
Simvastatin 40mg od

3   Explanation   **Patient role**

You are a 56 year old who is normally fit and well. You were not on any tablets before you came into hospital.

Last week you had a heart attack, so you have had treatment and been in hospital for a few days. You are ready to go home now.

You have seen the cardiac rehabilitation nurse already so most of your questions have been answered, but the nurse was called away before she could go over your medicines with you. She said that she would send the young doctor has come to discuss them.

Your medicines on the list are:

Aspirin 75mg od
Bisoprolol 5mg od
Clopidogrel 75mg od
GTN spray prn
Ramipril 10mg od
Simvastatin 40mg od

You will want to know what the tablets are for, and when to take them all.

It is difficult to remember so many medicines. Will you have to be on them for ever?

You are not sure that you like taking tablets all the time, but then again you do not want to have another heart attack.

How will you know if the tablets are working?

3 Explanation

Examiner mark sheet

Introduces self and identifies role

Checks patient's identity

Confirms/ establishes reason for talk

Establishes patient's understanding of the situation

Facts to include in the discussion:

Aspirin

- thins the blood a little to reduce the chance of another heart attack (because heart attacks are caused by clots blocking the arteries).
- Taken once per day
- Usually no side effects but tell the doctor if you get indigestion
- Avoid taking tablets such as ibuprofen (nurofen) with aspirin as they can interfere with it and cause indigestion

Bisoprolol

- Slows the heart a little to help it work better, lowers blood pressure.
- Taken once per day

Clopidogrel

- Also thins the blood a little.
- Taken once per day

GTN spray prn

- Use this if you get angina (chest pain)
- Should relieve the pain within five minutes, if not use once more but if this does not work then call 999
- Sit down when you use it as it can cause dizziness and headache.

Ramipril

- Helps control the blood pressure; may help the heart muscle heal.
- You will need regular blood tests to make sure the tablet is agreeing with you (it can affect the kidneys).

Simvastatin

- Lowers the cholesterol to reduce the chance of more heart attacks.
- Take one at bedtime.

Advises that if the patient has any side effects from any of the medicines to let the doctor know, so that the treatment can be adjusted.

Uses chunking and checking - giving small pieces of information and checking understanding before continuing

Establishes patient's concerns, validates and addresses these (remembering to take the tablets; duration of treatment)

Checks if patient has other questions

Uses communication tools, such as repetition and summarising, appropriately

Offers written advice or websites

Language appropriate throughout

Friendly approach, appropriate body language

Fluent and professional manner

3 Notes

See NICE guidance at www.nice.org/CG48

The most difficult aspect to this station is knowing how much or how little information to give. Make sure you keep on checking with the patient – are they following you? Have you answered the questions they have?

You should know the common indications, contra-indications, side effects and necessary monitoring for the most widely used drugs, but don't tell the patient everything you know. You could include information about cough as a potential side effect of ACE inhibitors, and about muscle aches from statins but the most important point during this consultation is to address the patient's concerns, not to say everything.

> LEARNING POINT
> Make yourself a personal formulary of drugs you have met, to help you become familiar with the most commonly prescribed medicines. Use a table such as this:

Drug (name & class)	Mechanism of action (What is does)	Indications (what it is given for)	Contra-indications and side effects	Monitoring needed (if any)	When to take it/ special instructions to patient
Aspirin (anti-platelet, anti-pyretic, anti-inflammatory)	Inhibits COX to reduce thromboxane synthesis and so make platelets less sticky	Ischaemic heart disease, Atrial fibrillation, Post-thrombotic stroke	Can cause indigestion, danger of Reye's syndrome in children, Hypoglycaemia and tinnitus in overdose	None	Usually taken with food in the morning
Bisoprolol (beta-blocker)	Slows the heart by blocking beta adrenoreceptors	Post MI Angina Hypertension CCF Anxiety	Contraindicated in asthmatics Contraindicated in PVD, Causes cold hands and feet, impotence, tiredness	Pulse and BP, signs of cardiac failure	Nil special
Ramipril	Lowers the blood pressure by...	Hypertension CCF ...	...		

4 History   **Candidate role**

You are a foundation doctor in a respiratory ward.

A 53 year old has been admitted from Accident and Emergency for further investigation of shortness of breath.

Please take a history.

When you have finished tell the examiner your differential diagnosis (most likely first)

4 History   Patient role

You are 53 years old and have been sent up to a ward from Accident and Emergency, which you came to three hours ago.

You have been having problems breathing for a couple of months – this has got steadily worse. You first began to notice you had difficulty in breathing when climbing stairs. Last week you began to get short of breath when walking more than a couple of streets and last night the breathlessness woke you up and you had to get up to 'get some air'. You did not get back to sleep and the feeling did not go away. You are really quite frightened, you hate hospitals, as you feel the staff never like homeless people, or treat you with respect.

Background

You have your usual smokers cough. You cough up a bit of phlegm in the mornings. It is white in colour – sometimes it goes yellow. You have never seen any blood. You have no chest pain or discomfort, back pain, problems with your waterworks, or abdominal pain.

You do get swollen ankles by the end of the day but they get better by the morning if you get a good night in a bed , if not they may stay up for a few days till you get some proper rest.

You were diagnosed as having high blood pressure about 15 years ago and told you have diabetes a few years after that. You have been given various medicines (all tablets) over the years for your hypertension and diabetes but rarely collect the prescriptions as you are not keen on taking tablets. Why take medicine if you do not feel ill?

You are currently unemployed and homeless, staying in a local hostel. You have no GP, but attend the A & E department for small health problems and occasionally have been brought in after you have had drinking binges.

You smoke cigarettes and drink alcohol frequently (when you have money), but can't really say how much. You have no idea if you are allergic to anything.

You have not seen anyone from your close family for many years and do not know whether any of them are still alive. Your mother had an operation to remove a swelling from her neck when you were young.

4   History   **Examiner mark sheet**

Introduces self and role and checks patient identity
Explains purpose of interview
Establishes nature of presenting complaint
- time duration
- exacerbating and relieving factors
- severity
- previous episodes/ how is he normally?

Enquires about relevant associated symptoms:
Respiratory
- cough
- sputum
- haemoptysis
- wheeze
- exercise tolerance

Cardiac
- chest pain
- palpitations
- ankle swelling
- orthopnoea
- paroxysmal nocturnal dyspnoea

Establishes effect on daily life
Establishes patient's concerns, validates and addresses these (fear of hospitals)
Drug history (including over the counter and illicit drugs)
Allergies
Establishes previous medical history
Establishes family medical history
Establishes Social history
- occupation
- who is at home
- smoking
- alcohol

Excludes other systemic symptoms
Appropriate questioning technique (mixture of open and closed questions)
Avoids or explains jargon
Uses tools such as signposting and summarising
Systematic, logical approach
Checks if the patient has any other questions
Finishes consultation in suitable manner
Friendly approach, appropriate body language

4 Notes

Differential diagnosis

- Congestive cardiac failure – with hypertension and diabetes mellitus as risk factors. The symptoms are fairly typical in this case (including ankle oedema and paroxysmal nocturnal dyspnoea).
- Chronic obstructive pulmonary disease –is clearly in the background but no exacerbation as sputum is white
- Lung cancer – patient is a smoker
- Other lung diseases – fibrosis – did you take a history of past occupation?

5 History   Candidate role

You are a foundation doctor in General Practice. Your next patient is aged 72 years. S/he has come complaining of palpitations. Please take a history.

At the end of the history, please
- summarise your history to the examiner
- state the possible causes of the symptoms described
- tell the examiner the investigations you would like to carry out

5 History   **Patient role**

You are 72 years old.

You have come to the doctor today because you have been getting 'funny do's' where you could feel your heart beating fast and you can't catch your breath.

You are quite worried that this may be a heart attack, and worried that all this might involve going into hospital – you don't like hospitals because they are noisy places and there is no privacy.

These 'funny do's' have happened about four times now – the first time was last Thursday when you were walking to the shops, then when you were lying in bed a couple of days later. It also happened twice over the weekend when you were just sitting down reading the paper. Each time it happened, you just waited and the feeling passed off by itself after a few minutes. There was no chest pain or dizziness. You couldn't really say if the heartbeat was regular or irregular – it just felt really fast all of a sudden.

Background

You have had high blood pressure for many years – you take amoldapain and bendro-something. According to your regular GP your blood pressure is 'not bad'. About 20 years ago you had a hernia operation. Apart from that you keep very well normally.

You are a retired bus conductor.

You live with your spouse, who is generally quite well, though s/he has arthritis and problems with the thyroid. You live in a ground floor flat, which makes it easier for your spouse to get about. You look after yourselves and can get to the shops and out to visit friends.

Your children are grown up – your eldest son works for an engineering firm and is Dubai at the moment. Your daughter is a housewife and lives not far away. They have not had any significant health problems.

You used to smoke 20 cigarettes a day for about thirty years. You gave up in your 50's when you were diagnosed with high blood pressure. You like a glass or two of beer at the weekend, and an occasional whisky at home (once or twice a week).

You have no other problems, except for a tendency to get constipated (which has happened on and off over the last few years). You buy laxatives from the chemist to help with that.

5 History   **Examiner mark sheet**

Introduces self and role and checks patient's identity

Explains purpose of interview

Establishes nature of presenting complaint

- onset
- character (rhythm)
- time duration
- exacerbating and relieving factors

Enquires about relevant associated symptoms:

Related to primarily cardiac causes

- Chest pain
- Breathlessness
- Nausea
- Sweating
- Ankle swelling
- Claudication
- Cough
- Dizziness
- Weight loss

Establishes effect on daily life

Establishes patient's concerns, validates and addresses these (fear of going into hospital)

Drug history (including over the counter and illicit drugs)

Allergies

Establishes previous medical history

Establishes family medical history

Establishes Social history

- occupation
- who is at home
- smoking
- alcohol

Excludes other systemic symptoms

Appropriate questioning technique (mixture of open and closed questions)

Avoids or explains jargon

Uses tools such as signposting and summarising

Systematic, logical approach

Checks if the patient has any other questions

Friendly approach, appropriate body language

Finishes consultation appropriately

5 Notes

Likely causes

Palpitations have a number of causes, including

- cardiac disease
- medication
- thyroid disease
- anaemia
- anxiety
- caffeine
- alcohol/ drugs of misuse

Investigations

Resting ECG
24 hour/ ambulatory ECG
Full blood count
Thyroid function
Fasting lipids/ glucose

LEARNING POINT

Summarising

Learning to summarise a history and present the findings to a senior colleague is a very important skill. It can be frustrating at first as you will find that each doctor you present to seems to be wanting a different level of detail. As a general rule, if you can summarise in two to three sentences, you will please most people. You can always add more detail if asked.

Try to include the most important findings in the history, including key negatives as well as positives and relevant points of the past medical history, which may point you towards a diagnosis. In this instance, a useful format would go something like this:

'Mr/s X is a 72 year old wo/man with a past history of hypertension, who presents today following several episodes of palpitations, lasting a few minutes each time. These have come on during exercise and at rest and are associated with the feeling of breathlessness but no pain. She cannot say whether her heartbeat is regular or irregular during these episodes.'

---

6 Explanation   Candidate role

You are a Foundation doctor working in General Practice.

Your next patient is a 52 year old (BMI 24) who has essential hypertension. S/he has been on enalapril 20 mg for three months, but her blood pressure has remained high (usually around 160/100 mmHg). The clinic nurse has sent the patient to see you because at the check today it was 164/102 mmHg.

Please discuss the situation with the patient and reach an agreement about future management.

6 Explanation   Patient role

You are 52 years old with high blood pressure, diagnosed six months ago. You have been prescribed tablets for about four months now. Now the clinic nurse has asked you to see the doctor because your blood pressure is still high (164/102 mmHg today) and has never really come down less than 160 since it was diagnosed.

You haven't told the doctor but actually you stopped taking the tablets after the first month as you noticed that you were coughing more when you took them. You are asthmatic, so you don't want to take anything that upsets your chest. You will only admit this if the doctor really makes you feel comfortable. You hate being told off.

Background

You do not smoke – you gave it up years ago when cigarettes went up to more than a pound a packet. Your asthma has been better since giving up smoking too.

Your mother suffered from high blood pressure and diabetes. Eventually she had a stroke and now she is in a nursing home.

The only other medication you have is a blue reliever inhaler for your asthma. You only use this occasionally (once or twice a week – maybe a bit more often in the summer hay fever season).

You work as a health care assistant in the community. You get plenty of exercise walking every day, except when you really need to use the car for work. You are not particularly worried about having high blood pressure as you feel perfectly OK (except when you take those tablets).

6 Explanation   **Examiner mark sheet**

Introduces self and identifies role

Checks patient's identity

Confirms/ establishes reason for visit

Establishes understanding of the situation:

- the reason for treating high blood pressure is to reduce the risk of heart problems and stroke in the future, not because high blood pressure makes you feel ill at the time

Establishes patient's concerns, validates and addresses these

- concordance/ compliance and problems with side effects due to medication
- offers alternative medication that will not cause cough, following NICE guidance[4]
- agrees follow up

Checks if patient has other questions

Uses chunking and checking - giving small pieces of information and checking understanding before continuing

Uses communication tools, such as signposting and summarising, appropriately

Offers written advice or websites

Language appropriate throughout

Friendly approach, appropriate body language

Fluent and professional manner

---

[4] For relevant NICE Guidance see www.guidance.nice.org.uk/CG127

6 Notes

> LEARNING POINT
>
> If the medication is not working check your diagnosis and the patient's compliance.
>
> This case is included to remind you to check patient compliance when a drug does not seem to be working. In general practice you may be able to see from the computer how often the patient is collecting repeat prescriptions, which can give you a clue. But don't be fooled by the patient who collects medication regularly and stores up boxes of it at home, too afraid to tell the doctor that they are not taking it.
>
> Give the patient permission to admit they are not taking the tablets by using phrases such as
>
> 'I know some patients find taking tablets every day difficult. How do you get on?'

7 Explanation   Candidate role

You are a Foundation doctor in General Practice.

Your next patient is a 52 year old obese patient (BMI 31) who has essential hypertension (confirmed on ambulatory blood pressure monitoring). S/he has been trying to make lifestyle alterations following your advice on blood pressure, but the blood pressure remains high (usually around 160/100 mmHg). The nurse has been monitoring progress and has now asked the patient to see you for further help.

You consider that an angiotensin-converting enzyme inhibitor or angiotensin receptor blocker would be the best options. Please explain the medication and any future monitoring that will be necessary.

7 Explanation   **Patient role**

You are a 52 year old with high blood pressure.

When the high blood pressure was first diagnosed, you did not want to go on tablets and the doctor thought if you could lose some weight and do a bit more exercise it would help. You are aware that you weigh more than you should – the doctor has told you about it before, when you came in with arthritic knees. You have been doing your best but the blood pressure is still up. Now the clinic nurse has asked you to see the doctor again. You don't want another row about your weight – you have tried for years to diet and nothing has worked. It must be in your genes. Your mother suffered from high blood pressure and diabetes. Eventually she had a stroke and now she is in a nursing home.

Background

You do not smoke – you gave it up years ago when cigarettes went up to more than a pound a packet. Your asthma has been better since giving up smoking too.

You work as a health care assistant in the community, so you do get plenty of exercise walking up and down stairs and round the local estate to your clients.

You suppose tablets are inevitable now. At least that way you know you will not have a stroke like your Mum

7   Explanation   **Examiner mark sheet**

Introduces self and identifies role

Checks patient's identity

Confirms/ establishes reason for visit

Establishes understanding of high blood pressure and why it is treated

- to reduce the risk of heart problems or stroke later in life - no guarantees

Discusses new treatment is an ACE inhibitor or A2RB

- Start on low dose and build up over a few weeks
- Take it at bedtime as very rarely dropping the blood pressure with the first tablet may make a patient feel dizzy and would need to lie down
- Needs to have regular blood tests and BP checks - first checks within one to two weeks after starting new tablet
- Once the dose is correct for the individual, then checks will be every six months to a year
- Treatment is lifelong as there is no cure
- Lifestyle changes should continue.

Uses chunking and checking - giving small pieces of information and checking understanding before continuing

Checks for patient's concerns/questions

Offers written information or websites

Agrees follow up

Uses communication tools, such as signposting and summarising, appropriately

Language appropriate throughout (avoids or explains jargon)

Fluent and professional manner

Friendly approach with appropriate body language

7 Notes

> LEARNING POINTS
>
> When explaining new medication to patients, think about what they need to know
>
>> What it is
>>
>> What it is used for
>>
>> When to take it
>>
>> How long to take it for
>>
>> Any special monitoring needed
>>
>> Any side effects to look out for
>
> There are some side effects that you may wish to warn the patient about, for example when starting a patient on warfarin you will warn them to come straight to you if they get bleeding or unusual bruising. Other side effects you may choose not to mention but must be aware of so that if the patient presents with possible side effects you recognise the link to the medication. In the case of ACE inhibitors the commonest side effect to look out for is a dry cough.

8 Explanation   Candidate role

You are a Foundation doctor in general practice.

Your next patient is 56 years old. The patient has recently been discharged from hospital following treatment for an ST elevation MI and has come in to see you now for the first time after discharge.

Please talk to the patient, answering any questions and addressing any concerns.

Give appropriate lifestyle advice for a patient post-MI.

8   Explanation   Patient role

You are a 56 year old who is normally fit and well.

Last week you had a heart attack and spent several days in hospital. The doctors told you that they had cleared the blockage in the artery and you should make a good recovery. You are on a number of tablets but the doctor in hospital already explained those, so you don't need the doctor to go over that again.

You were told to go and see your GP when you went home, so now you have come to the surgery.

You have not had any pain since you came out of hospital. But you are still not entirely well... you feel like you have suddenly turned old overnight! You feel tired and quite anxious. You can't understand why this has happened to you as you always tried to keep quite fit. You go to the gym once or twice per week – you imagine that will be too much now.

You are concerned about how soon you can drive, as you want to get back to your job as a school bus driver.

Your children and spouse are very concerned about your health and you feel they are making too much fuss. They are making sure that you eat the right things now. The hospital told you to eat lots of fish. You are eating so much fish that you think you might swim away!

You smoked two cigarettes a day until last week, but you have given up now you have had the heart attack.

You might be wondering about your sex life – but you would be embarrassed to ask the doctor. You suppose that is the end of that. You wouldn't want to risk it.

8 Explanation   **Examiner mark sheet**

Introduces self and identifies role

Checks patient's identity

Confirms/ establishes reason for talk

Establishes patient's understanding of the situation

Facts to include in the discussion:

Check s patient understands medication

Explains importance of following rehabilitation team's advice:

- gradual, planned increase in activity, following the cardiac rehab team's advice
- plenty of rest
- Mediterranean diet (including oily fish 2 – 4 times per week)
- not smoking

Discusses return to activities of daily life

- Home life
- Driving - Explains the rules around driving after a heart attack or at least offers to check the rules around driving before patient goes home
- Work – Return to work depends on the job the patient does. In this instance the need for a passenger vehicle driver's license means the patient will have to inform the DVLA and follow their advice. It may be possible for the patient to return to work earlier if her employer can find her a non-driving role.
- Sex - Explains that normal sexual activity is OK once she can climb two flights of stairs without a rest

Establishes patient's concerns, validates and addresses these (why this happened despite being a fairly fit person; desire to return to work; sex life)

Explains timing and purpose of review as outpatient

Offers review if further problems or concerns

Uses chunking and checking - giving small pieces of information and checking understanding before continuing

Checks if patient has other questions

Uses communication tools, such as signposting and summarising, appropriately

Offers written advice or websites

Language appropriate throughout

Fluent and professional manner

8 Notes

> LEARNING POINT
> You should be aware of the conditions for which the Drivers Vehicle Licensing Authority impose special conditions on driving, such as heart conditions, neurological conditions, type 1 diabetes mellitus, sleep apnoea. You should always advise patients of the need to inform the DVLA of a change to their medical condition, and the need to refrain from driving if appropriate. If you do not learn the key restrictions at least remember to look up the advice.
> Current advice relating to patients after an acute coronary syndrome with no complications is that driving is not permitted for one week if successful angioplasty is used; one month after thrombolysis; longer for HGV or passenger vehicles
> See NICE guidance at www.nice.org/CG48 for management of patients post MI.

9   Explanation   **Candidate role**

You are the Foundation doctor in casualty.

You are going to see a middle aged patient with a history of angina. S/he has come in with severe chest pain, unrelieved by GTN. The ambulance crew gave them pain relief and have done an ECG. The ECG looks like an acute STEMI.

Please explain to him what you think has happened and what will happen now.

YOU DO NOT HAVE TO TAKE A HISTORY. THE DIAGNOSIS IS MI.

9   Explanation   Patient role

You are a 68 year old man who suffers with angina. Normally, this is relieved by your GTN spray, but this morning you woke very early with terrible chest pains, and your GTN didn't help. You called 999 as the pain was so bad. The ambulance men have given you an injection for the pain and done a tracing of your heart but they have not told you much yet.

Now you have arrived in hospital.

You want to know:

- Why didn't your GTN spray work?
- Is it a heart attack?
- What is the difference between angina and a heart attack?
- Are you are going to die?
- What will happen now?

9   Explanation   Examiner mark sheet

Introduces self and identifies role

Checks patient's identity

Confirms/establishes reason for hospital admission

Establishes understanding of result of ECG

Facts to include in the explanation:
- What the problem is
- What a heart attack is
- The difference between angina and a heart attack and why the GTN did not work this time
- What will be done about it

The treatment that will be either:
- a clot busting drug given via a drip to dissolve the clot in the arteries;
- or an attempt to clear the blockage in the artery by inserting a fine tube into the blood vessels in arm or groin and guiding it up to the arteries in the heart using X-ray monitoring

Establishes patient's concerns, validates and addresses these (difference between angina and MI and non-effectiveness of GTN spray; fear of dying)

Uses chunking and checking - giving small pieces of information and checking understanding before continuing

Reassuring without giving false guarantees

Checks if patient has other questions

Uses communication tools, such as signposting and summarising, appropriately

Language appropriate throughout (avoids or explains jargon)

Fluent and professional manner

Friendly approach, appropriate body language

9 Notes

LEARNING POINT

Look back at the tips in the introduction on dealing with frightened patients.

Watch how the doctors on the wards and in the clinics cope with anxious patients and reflect on what you think works well. The doctor's bedside manner is an important part of the treatment.

10 Explanation   Candidate role

You are a Foundation doctor in cardio thoracic surgery.

You are asked to speak to a patient's relative (son or daughter), who is visiting their father, a 68 year old man on your ward. The father was due to have his aortic aneurysm repaired this morning, but the operation has been cancelled because an earlier patient arrested on the operating table and eventually died. At the moment, you are not sure when the operation will be rescheduled for.

10 Explanation   Relative role

You are visiting your father, who is a 68 year old patient on a cardio-thoracic ward.

Your father was due to have his aortic aneurysm repaired this morning, but the operation has been cancelled. You are very worried, because you have been told that the aneurysm could kill your father if it bursts and that the longer it is left the more problems there will be. Your father really needs his operation.

You have been messed around by this hospital too long. You need to know what is going on. Your whole family has travelled from London to be with your father for this operation.

You ask to speak to one of the doctors to find out what is happening.

☐

10 Explanation   **Examiner mark sheet**

Introduces self and identifies role

Checks who they are speaking to - identity and relationship to patient

Confirms patient consent to talk with relatives (possibly by including patient in discussion)

Confirms/establishes reason for visit

Establishes concerns, validates and addresses these (worry over risks to father)

Facts to include in the explanation:
- acknowledges the distress caused
- apologises for the circumstances
- an unexpected emergency meant that the patient's operation has had to be delayed

Offers positive action
- the operation will be rescheduled as soon as possible – when the theatre, the surgeon and a high dependency bed are available
- Father will be monitored on the ward

If unable to calm the relative down offers senior to talk

Calms down situation without making promises that may not be kept

Checks for unanswered questions

Remains calm and polite throughout

Language appropriate (avoids or explains jargon)

Fluent and professional manner

Friendly approach with appropriate body language

10 Notes

Respiratory practice scenarios

11 History   Candidate role

You are a Foundation doctor in general practice.

Your patient is 49 years old and has come to see you about a chest problem. Please take a history.

When you have finished please tell the examiner your differential diagnosis (most likely first).

11 History   Patient role

You are 49 years old

You have had a cough for the past 10 days. The cough is generally dry, with occasional clear sputum but no blood. You have been slightly short of breath around the house and have felt dreadful for the past four days, like you've had flu. You have not noticed any wheezing. You have been feeling feverish in bed at night and your muscles and joints all hurt.

Background

You have had to have a couple of days off work. You work in IT and have a big deadline coming up.

You have not had any other big problems.

You smoke 20 cigarettes a day and work in the accounts department of a high street bank.

You have not been on any foreign travel recently and have not been in contact with anyone else with similar problems.

You are not on any regular medication and are otherwise well.

Only tell the doctor if you are specifically asked or if the doctor really gives you a chance to talk about your worries; two of your prize parrots have been very sick recently and one of them died three days ago. You had not really thought about any connection until now but that is a bit odd isn't it?

11 History   # Examiner mark sheet

Introduces self and explains role

Explains purpose of interview

Establishes nature of presenting complaint (cough)

- onset
- exacerbating and relieving factors
- sputum
- haemoptysis
- breathlessness
- wheeze
- chest pain

Enquires about relevant associated symptoms:

- fever
- palpitations

Establishes effect on daily life

Establishes patient's concerns, validates and addresses these (illness of parrots)

Drug history (including over the counter and illicit drugs)

Allergies

Establishes previous medical history

Establishes family medical history

Establishes Social history

- occupation
- who is at home – including pets!
- smoking
- alcohol

Excludes other systemic symptoms

Appropriate questioning technique (mixture of open and closed questions)

Avoids or explains jargon

Uses tools such as signposting and summarising

Systematic, logical approach

Checks if the patient has any other questions

Friendly approach, appropriate body language

Finishes consultation appropriately

11 Notes

Differential diagnosis

Chest infection – possibly Chlamydia psittica

LEARNING POINT

The sick birds give a clue to an unusual type of infection that may be present in this patient. Birds infected with Chlamydia psittica may pass the infection on the people. You may also think about avian flu. Birds can also be associated with a form of extrinsic allergic alveolitis. This is not an infection but a type 2 sensitivity reaction to inhaled bird proteins. It is sometimes known as bird (or pigeon) fancier's lung. It presents with prolonged insidious symptoms of breathlessness and vague generalized symptoms of low grade fever, malaise and myalgia, rather than a cough.

This station really tests your ability to let the patient give you the answer.

SUGGESTED ACTIVITY

Do you know how to test for atypical causes of pneumonia such as C. Psittica? Chlamydia is an intracellular organism, so is difficult to culture. Find out how to test for it in your area.

You might also like to revise other causes of atypical pneumonia, such as

Legionella pneumophila

Pneumocystis jevicii

Mycoplasma pneumonia

Coxiella burnetti

---

12 History   Candidate role

You are a Foundation doctor in casualty.

A 61 year old patient has been admitted with breathlessness and chest pain. An ECG was taken, but was grossly normal, with no evidence of ischaemia. Please take a history and discuss the possible diagnoses and next steps with the patient.

12 History   Patient role

You are 61 years old.

You have been admitted to accident and emergency because you were breathless this morning and had chest pain. The pain was in the right side of your chest and sharp. It did not go anywhere else. Nothing made the pain better until you came into hospital and the nurse gave you an injection. You had that heart tracing done, but the doctor said it didn't look like there was a problem with your heart.

Background

You are a retired car factory worker.

You are a lifelong smoker, smoking 30 cigarettes a day. You have been getting progressively unwell over the last 2-3 years. Every winter for the last 8-9 years, you have had a severe cough with thick green sputum. You have never noticed any blood in my sputum. This has worsened and now seems to happen all the year round.

You can only walk about 50 - 60 metres on the flat without stopping, due to breathlessness and wheeze. You have a near constant cough and have to sleep with three pillows otherwise you are more breathless.

You have not been feverish.

You do not usually suffer with any chest pains, swelling of the ankles or episodes of breathlessness at night.

Your GP recently started you on some inhalers but you rarely use them as they seem to do little good. You take no other regular medications and you are otherwise well, with no other major problems.

You are feeling worried that your lungs are never going to get better. You don't want to become a burden on your family as they already have a lot on their plates. Your mother and your father both have dementia.

12 History   Examiner mark sheet

Introduces self and role and checks patient identity
Explains/ confirms purpose of interview
Establishes nature of presenting complaint
- site
- onset
- character
- radiation
- exacerbating and relieving factors
- time duration
- severity

Enquires about relevant associated symptoms:
Respiratory
- cough
- sputum
- haemoptysis
- breathlessness
- wheeze
- chest pain

Cardiac
- Palpitations

Establishes effect on daily life
Establishes patient's concerns, validates and addresses these (deterioration; burden on family)
Drug history (including over the counter and illicit drugs)
Allergies
Establishes previous medical history
Establishes family medical history
Establishes Social history
- occupation
- who is at home
- smoking
- alcohol

Excludes other systemic symptoms
Appropriate questioning technique (mixture of open and closed questions)
Avoids or explains jargon
Uses tools such as signposting and summarising
Systematic, logical approach
Checks if the patient has any other questions
Friendly approach, appropriate body language
Finishes consultation appropriately

12 Notes

Differential diagnosis

Chronic obstructive pulmonary disease as a cause of breathlessness with another pathology causing the pain, such as

    infective exacerbation with an element of pleurisy

    lung cancer eroding a rib

    rib fracture (possibly due to metastatic disease).

13 Explanation   Candidate role

You are a foundation doctor in general practice.

Your next patient is a 65 year old who you have been treating for a chest infection. The patient seemed to improve after a course of amoxicillin, but never got quite back to normal. S/he remained quite short of breath, so you requested a chest x-ray.

The patient is a lifelong smoker.

Now you have the X-ray result. It shows an opacity in the left upper lobe. And the radiologist suggested urgent referral for further investigation by the respiratory physicians.

Please discuss this result and the referral with the patient.

13 Explanation   Patient role

You are a 65 year old retired sales representative.

You have a bit of a cough that has been bothering you ever since you caught a cold a couple of months ago. The GP gave you some antibiotics and it got a bit better, but you were still a bit breathless so you had to go back and see him again. He sent you for a chest x-ray and today you have come in to get the result.

You have smoked since the age of 14 and are starting to get just a bit worried (that you might have lung cancer). Your daughters have been on at you to stop smoking and you have seen all the health warnings on the packet. You are trying not to think about the possibilities…but it does play on your mind.

Background

You are not on any medication. The only major illness you had in the past was having your gallbladder removed aged 45.

You don't know how your other half would manage if anything were to happen. You don't want them to worry. You have always been the one who does the repairs and looks after things around the house.

13 Explanation   Examiner mark sheet

Introduces self and identifies role

Checks patient's identity

Confirms/ establishes reason for visit

Establishes understanding of test and of the expected result

Asks how patient is feeling now

Facts to include in the explanation:

What the result shows

- The X-ray shows a shadow in the left lung

What it means

- More tests are needed to find out the cause of the shadow
- There are a number of possible causes
- It is possible that this is cancer, but more tests are needed before anyone can know for sure

What will happen now

- The patient will be referred to see a chest specialist within two weeks
- The specialist will decide what other tests are needed
- It is likely that these will include more scans and other tests

Begins gradually and gauges patient's reaction, taking time to allow for response

Uses chunking and checking - giving small pieces of information and checking understanding before continuing

Establishes patient's concerns, validates and addresses these (lung cancer; supporting wife)

Checks if patient has other questions

Uses communication tools, such as signposting and summarising, appropriately

Offers written advice or websites

Language appropriate throughout

Friendly approach, appropriate body language

Fluent and professional manner

13 Notes

> LEARNING POINT
>
> This station is a difficult one. You will need to begin gently and break the news gradually. It is important that you do not deny the possibility that this X-ray lesion could be cancer, as this would be untruthful and the patient would lose trust in you. Be clear to the patient that you still do not know the definite diagnosis and that there are other possibilities (such as infection).
>
> Under current guidance in the NHS, suspected cancer (such as this) must be referred to an appropriate specialist to be seen within 2 weeks. There is little evidence that this improves patient survival, but it is clearly important in terms of reducing the time patients are left worrying.

14 History and explanation   Candidate role

You are a foundation doctor in general practice.

Your next patient is a twenty-three year old who has recently joined your practice.

You see from the computer that the patient is normally fit and well and on no medication.

Please take a history and deal with the problem presenting today.

If you need to examine the patient or to do investigations, ask for the findings when you are ready.

14 History and explanation   Patient role

You are aged twenty three.

You are normally fit and well.

You are not on any medicines. You smoke ten cigarettes per day.

You work in a garage and have had the morning off to come to the doctor's as you have had a dry tickly cough for the last five days. You have had no other symptoms.

You think that antibiotics will help as your old GP always used to give you them for things like this. You are going to Majorca on Monday and want to be better for then.

If the doctor wants to examine you, you can tell him that the examination is normal.

Investigations are not needed for this station. If the doctor chooses to do investigations tell him that the results will take four days to be reported.

14 History and explanation   Examiner mark sheet

Introduces self and role and checks patient identity

Explains purpose of interview

Establishes nature of presenting complaint (as relevant)

- onset
- sputum
- haemoptysis
- exacerbating and relieving factors

Enquires about relevant associated symptoms:

- breathlessness
- wheeze
- chest pain
- fever
- ear symptoms
- sore throat

Establishes effect on daily life

Establishes patient's concerns, validates and addresses these (need for antibiotics; imminent travel)

Drug history and allergies

Establishes previous medical history

Establishes Social history

- occupation /who is at home
- smoking and alcohol

Facts to include in explanation:

What findings show

- History and examination do not suggest a serious underlying illness, only an upper respiratory infection

What this means

- Explains lack of evidence for antibiotics with potential harmful effects

What can be done now

- Advises simple home remedies, such as honey and lemon, paracetamol, steam inhalation

Handles situation diplomatically

Appropriate questioning technique (mixture of open and closed questions)

Avoids or explains jargon

Uses tools such as signposting and summarising

Systematic, logical approach

Checks if the patient has any other questions

Friendly approach, appropriate body language

Finishes consultation appropriately

14 Notes

LEARNING POINT
This station illustrates some important skills you will need as a doctor.
These include the ability to:
> Recognise minor self-limiting illness that can be managed using simple home remedies.
> Challenge patient's health beliefs
> Modify health seeking behaviour
> Defuse a potentially confrontational situation

15 Consent   Candidate role

You are a foundation doctor in the Respiratory clinic. Your next patient is a 69 years old, booked for a bronchoscopy for suspected carcinoma of the lung. Please consent the patient for the bronchoscopy.

15 Consent   Patient role

You are 69 years old.

You have come to the clinic to see the lung specialist and have some tests.

You have been breathless and have coughed up blood a couple of times.

You suspect it is lung cancer as you have always been a heavy smoker and your GP agreed that it was a possibility.

One of the doctors is going to explain a 'camera test' to you. You want to know whether they will be able to tell you the result later today. Also, a friend said you would watch a television showing your insides as the camera went down. You don't want to watch and want to ask the doctor if it is essential to have the telly on.

15 Consent   **Examiner mark sheet**

Introduces self and identifies role

Checks patient's identity

Confirms/ establishes reason for visit

Establishes understanding of test and of expected result

Facts to include in the explanation:

What will happen (and why)

- the procedure is intended to look inside the airways using a camera on a long, thin tube
- the tube is about the thickness of a little finger (or stethoscope)
- the patient will be given a little sedation to make them sleepy, but will still be awake
- the patient will be connected to a heart monitor with wires and sticky pads on their shoulders and left side
- the doctor will numb the inside of their mouth (and possibly the nose) with an anaesthetic spray
- the doctor will then pass the tube through their mouth or nose and look around the airways using the camera
- they may take a very small sample of lung tissue to have it checked in the laboratory – this will not hurt as the lungs themselves do not have nerve endings for pain
- the bronchoscopy takes about 20 - 30 minutes

After the procedure

- the patient can go home later that day once properly awake
- the patient will need someone to collect them and take them home and stay overnight, in case of problems
- no driving or signing important forms for 24 hours afterwards
- an appointment will be made for the patient to come back to clinic to get the results once the laboratory has had a chance to look at the samples (usually after a couple of weeks)

Minor, common complications - there may be some discomfort in the back of the throat afterwards, but the patient will be given painkillers

Rare, serious complications - very, very small chance of damage to the lungs

Uses chunking and checking - giving small pieces of information and checking understanding before continuing

Establishes patient's concerns, validates and addresses these (DO NOT DENY IT MIGHT BE CANCER BUT STRESS THAT THE TEST RESULTS ARE NEEDED BEFORE YOU CAN KNOW). If the patient prefers not to look at the screen the nurse will usually help distract them.

Checks if patient has other questions

Uses communication tools, e.g. signposting and summarising, appropriately

Language appropriate throughout

Fluent and professional manner

Friendly approach, appropriate body language

15 Notes

> LEARNING POINT
> Discussing risk whilst consenting
> You should always discuss common minor risks and the more serious but rarer risks when consenting a patient for a medical procedure or operation. You do not have to know the exact figures relating to risks, but if the patient asks for them you should be prepared to go away and find out.
> Most hospitals have written information about the key procedures and operations, which state figures in them, so these can be useful to both doctor and patient.

Gastroenterology practice scenarios

16 History   Candidate role

You are a Foundation doctor in General Practice.

Your next patient is 26 years old, and has come to you to discuss a history of diarrhoea. The patient looks quite pale.

Please take a history

- At the end, please ask the examiner for any examination findings
- Then tell the examiner the investigations you would like to do
- Finally list your differential diagnoses (most likely first)

16 History   **Patient role**

You are 26 years old.

You have come to see the doctor today because you are having diarrhoea with blood in it. It is making work really difficult as you keep having to nip off to the loo all the time.

The diarrhoea started about two months ago. At first, it wasn't so bad and you thought it was just something you had eaten that had disagreed with you. You were having diarrhoea once or twice a day. For the last week it has been getting worse – you are getting stomach cramps with the diarrhoea, and you need to go to the toilet quite quickly. There has been fresh blood in the diarrhoea for the last two days. You are also starting to feel generally unwell, with joint pains and feeling a bit sweaty. The diarrhoea comes on four or five times a day now.

Your appetite has not changed, but you have been trying to eat less fruit and veg. to see if that would help (it hasn't).

You are worried that the diarrhoea hasn't settled down and you are going to get worse and worse. You have no idea what it might be.

Background

You work as a computer technician.

You are normally fit and well and have no problems other than the ones mentioned above. The only time you have been in hospital was when you broke a bone in your foot playing football as a teenager at school. You are not on any medication. You do not smoke, except an occasional joint. You usually drink about four or five beers on Friday and Saturday evenings.

You parents and two younger sisters are generally well. You have an aunt who has severe rheumatoid arthritis.

You went on holiday to Thailand this year. This was four months ago. NB You will only mention this if asked directly.

16 History   **Examiner mark sheet**

Introduces self and role and checks identity of patient

Confirms/explains purpose of interview

Establishes nature of presenting complaint (as relevant)

- onset
- character (colour, consistency, blood and/or mucus in stool)
- frequency
- exacerbating and relieving factors
- past experiences
- infectious contacts
- recent travel

Enquires about relevant associated symptoms:

- abdominal pain
- fever
- change in appetite
- nausea and vomiting
- weight loss
- joint pain
- skin problems
- eye problems

Establishes effect on daily life

Establishes patient's concerns, validates and addresses these (concern that problem will not get better)

Drug history (including over the counter and illicit drugs)

Allergies

Establishes previous medical history

Establishes family medical history

Establishes Social history

- occupation
- who is at home
- smoking
- alcohol

Excludes other systemic symptoms

Appropriate questioning technique (mixture of open and closed questions)

Avoids or explains jargon

Uses tools such as signposting and summarising

Systematic, logical approach

Checks if the patient has any other questions

Friendly approach, appropriate body language

Finishes consultation appropriately

16 Notes

Examination findings to give to candidate:

Pulse   92 bpm
BP   136/84
Abdo   generally tender, but not obviously distended
PR   Soft brown stool on glove with a trace of blood, no masses felt

Investigations:

Stool culture, including ova, cysts and parasites
Full blood count (anaemia), LFTs, ESR or CRP
Referral to specialist for sigmoidoscopy and biopsy

Differential diagnoses

Inflammatory bowel disease
Infectious gastroenteritis

LEARNING POINT
Never forget the importance of considering a travel history in patients presenting with what could be infectious symptoms.
Also remember that stool sent off for culture will only be cultured for pathogenic bacteria and not be examined for gut parasites (such as those found in the tropics) unless you specifically request this. Equally if you suspect a toxin producing bacteria such as Clostridium difficile, you will have to ask the lab to look for it.

17 History   Candidate role

You are a Foundation doctor in General Practice.

Your next patient is 55 years old woman. She is complaining of bleeding per rectum.

Please take a history. When you have completed your history the examiner will show you the examination findings. You must then

- state your differential diagnosis (most likely first)
- and formulate a management plan.

17 History   Patient role

You are a 55 year old woman.

You have seen spots of bright red blood in the toilet pan several times over the past month or so, only after you have passed a stool. Apart from that, you are well and take no medication. You have no pain and are eating normally. You find it difficult to keep your weight down. Your bowels are regular (you open them once a day, usually) and your motions have not changed in frequency or consistency.

Background

You have no longstanding health problems.

You had your appendix out when you were a teenager.

You live alone – your husband died in a car accident 20 years ago and your two daughters both live in a nearby town.

Your parents both died of cancer (lung and bone) in their sixties.

Your daughters were both born normally with no particular problems.

You went through the menopause two or three years ago. At the time you took Remifemin (Black Cohosh) to help with the hot flushes, but did not tell your GP because she was sceptical about herbal medicine. You are now worried in case the Remifemin has caused this problem – you read on the internet that it can cause thickening of the womb which might lead to cancer of the womb and maybe it is affecting your bowel as well? You are very worried about what might be wrong with you. Maybe it is cancer? Your parents died of cancer, so perhaps it runs in the family?

You work as a hotel receptionist. You enjoy a glass of wine most evenings and occasionally a cocktail or two at the weekend. You do not smoke.

17 History   Examiner mark sheet

Introduces self and role and checks identity of patient

Explains/ confirms purpose of interview

Establishes nature of presenting complaint (as relevant)

- onset
- character – mixed in with stool/on paper/in toilet/definitely pr
- amount
- frequency
- past episodes

Enquires about relevant associated symptoms:

- change in appetite?
- change in bowel habit?
- abdominal pain?
- weight changes?

Establishes effect on daily life

Establishes patient's concerns, validates and addresses these (is it cancer? Have I caused it?)

Drug history (including over the counter and illicit drugs)

Allergies

Establishes previous medical history

Establishes family medical history

Establishes Social history

- occupation
- who is at home
- smoking
- alcohol

Excludes other systemic symptoms

Appropriate questioning technique (mixture of open and closed questions)

Avoids or explains jargon

Uses tools such as signposting and summarising

Systematic, logical approach

Checks if the patient has any other questions

Friendly approach, appropriate body language

Finishes consultation appropriately

17 Notes

Examination findings

Patient appears well, no pallor, not underweight
Abdominal examination entirely normal – soft, non-tender, no masses, bs present
PR – no masses felt, soft brown stool on glove, no blood or mucus

Differential diagnosis

Piles (most likely)
Colorectal cancer (most important to exclude)
Rectal polyp

Investigations

FBC – to look for anaemia
CRP – as a general marker of disease
LFTs – worth checking liver function if you are thinking of possible cancer
Sigmoidoscopy – the definitive test here to exclude a malignancy in the lower large bowel

---

LEARNING POINT
You should be familiar with the criteria for urgent referral of cases which have symptoms or signs suggestive of cancer. The current target for these patients to be seen by a specialist in the NHS is two weeks – hence they are often referred to as two week rule referrals. See
http://www.nice.org.uk/CG027quickrefguide
to revise the red flag symptoms that necessitate an urgent referral under the two week rule.
Note that the two week rule is in fact an arbitrary target. There is little evidence that it makes a difference to the ultimate outcome if all patients are seen faster (say within one week) or slightly slower (say within three weeks).
Also be aware that some patients do not fall within the strict criteria of a two week referral, but would still raise your suspicion is concerning the possibility of cancer and should still be referred urgently. This patient is a good example of this. Although she does not quite fit the 2 week rule urgent cancer referral criteria, she still needs investigation. She may well have piles but would need to at least sigmoidoscopy to rule out a more serious underlying pathology.

---

18 History   Candidate role

You are a Foundation doctor in General Practice.

Your next patient is a 42 year old woman, who is complaining of abdominal pain.

Please take a history.

- At the end of the history, please tell the examiner your differential diagnoses
- State the investigations you would like to request.

18 History   Patient role

You are a 42 year old woman.

You have had quite bad stomach ache for a couple of weeks. It is on your right side, just under your ribs. It comes and goes on its own – sometimes it lasts for a few minutes, sometimes for an hour or more. Sometimes you get a mild pain there after eating. There is nothing else that particularly makes it worse or better. You have been taking paracetamol for the pain, which helps. The pain feels like a dull ache.

At first you thought you might have pulled a muscle or something, but it does not seem to be going away. You are worried that it might be a stomach ulcer.

You have not noticed any change in your bowels or your waterworks. You have not had any diarrhoea. You have not had any other health problems with this pain.

Background

You do not take any medication and have no allergies that you know of. You have no other health problems and you have never been in hospital.

Your periods are normally regular – every 31 days you bleed a small amount for 4 days. The last period you had was 2 weeks ago and was normal. You have two children, a boy and a girl, both teenagers at school. They and your husband are all well.

Your parents are both alive. Your father is 74, has high blood pressure and is disabled from a stroke – he has a weak left leg and needs a wheelchair if he goes out. He is looked after by your older sister. Your mother separated from him 20 odd years ago and lives in Jamaica. She has had a knee replacement for arthritis, but is otherwise well.

You work as a campaign assistant for an MP and are concerned that this pain may interfere with work as there is an election coming up. It is your busiest time. You drink the occasional glass of wine and do not smoke. You have not been abroad recently.

18 History   **Examiner mark sheet**

Introduces self and role and checks patient identity
Explains/ confirms purpose of interview
Establishes nature of presenting complaint (as relevant)
- site
- onset
- character
- radiation
- time duration
- exacerbating and relieving factors
- severity

Enquires about relevant associated symptoms:
- General: weight loss?
- Bowel:
  o change in appetite?
  o change in bowel habit?
  o stool (colour? diarrhoea? blood ?)
- Urinary symptoms?
- Gynaecological:
  o LMP and normal cycle
  o vaginal discharge
  o contraception/ chance of pregnancy

Establishes effect on daily life
Establishes patient's concerns, validates and addresses these (stomach ulcer; pain interfering with work)
Drug history (including over the counter and illicit drugs)
Allergies
Establishes previous medical history
Establishes family medical history
Establishes Social history
- occupation
- who is at home
- smoking
- alcohol

Excludes other systemic symptoms
Appropriate questioning technique (mixture of open and closed questions)
Avoids or explains jargon
Uses tools such as signposting and summarising
Systematic, logical approach
Checks if the patient has any other questions
Friendly approach, appropriate body language
Finishes consultation appropriately

18 Notes

Differential Diagnosis

Biliary colic
Peptic ulcer disease
Mild gastroenteritis

Investigations

FBC (raised white cell count in the presence of acute infection of cholecystitis)
LFTs (looking for raised bilirubin and raised Alk phos in obstruction)
Consider H pylori for peptic ulcer disease
Ultrasound scan – looking for gallstones

> LEARNING POINT
> Although this patient fits the classic description of patients with gallstones, (fair female fertile fat and forty) remember that most cases of gallstones occur in patients who do not fit this description. Also many cases of gallstones are found incidentally, up to 20% of people in the West have them and not all cause pain, so there is no guarantee that removing this patient's gallstones will resolve her problem.
> Be sure you know the different modalities of treating gallstones compared to kidney stones and avoid confusing the two.

19 Explanation   Candidate role

You are a Foundation doctor in General Practice. Your next patient (aged 25) saw the GP three weeks ago with typical symptoms of dyspepsia. The GP treated with a proton pump inhibitor (omeprazole 20mg) and asked the patient to have a blood test. The patient has now returned for the result of the blood test. It shows H. pylori serology positive. Please discuss the results with the patient.

19 Explanation   Patient role

You are 25 years old.

You saw your GP a few weeks ago because you had been having indigestion on and off for a couple of months. You had tried antacids, which helped but it still kept coming back. The doctor gave you a tablet to take every day and took a blood test. Now you have come back for the blood test result. You were a bit worried when the receptionist told you to make an appointment for the result – does that mean it is something serious?

If the doctor suggests further treatment you may be a bit reluctant to take more tablets. Actually you are feeling better since you started that omzole whatsit. You took it every day for the first couple of weeks but now you just take it if you have indigestion.

You are getting married next week, so if the doctor suggests antibiotics you will want to know if it is OK to drink with them.

You work in an insurance office. You drink a few glasses of wine/ beer at the weekends (but you will be drinking more for your wedding!). You don't smoke.

19 Explanation   Examiner mark sheet

Introduces self and identifies role

Checks patient's identity

Confirms/ establishes reason for visit

Establishes understanding of test and of the expected result

Facts to include in the explanation:

   What the result is / what the result means

- Sometimes indigestion can be caused by bacteria in the stomach, called Helicobacter.
- The blood test shows that the patient has helicobacter.

   What can be done about it

- It is easily treated with tablets.
- In most but not all patients the symptoms of recurrent indigestion settle after treatment.
- A few people still have indigestion symptoms after treatment and need tablets long term or to use when they get a flare up of the pain.
- Treatment consists of two antibiotics and a high dose of omeprazole (or other proton pump inhibitor), all taken together over a week.
- It is quite common for people to have a bit of an upset stomach (nausea and loose stools) on this treatment but it is important to complete it if possible.
- After finishing the treatment it is best to continue the omeprazole for four weeks more before stopping finally.

Establishes patient's concerns, validates and addresses these (is problem serious? drinking while on antibiotics):   It may be worth delaying treatment to avoid the wedding

Uses chunking and checking - giving small pieces of information and checking understanding before continuing

Checks if patient has other questions

Uses communication tools, such as signposting and summarising, appropriately

Offers written advice or websites

Language appropriate throughout

Friendly approach, appropriate body language

Fluent and professional manner

19 Notes

LEARNING POINT

Additional facts to know, but probably not to include in your explanation unless in response to patient questions:

First line treatment of helicobacter is usually with amoxicillin, clarithromycin and a proton pump inhibitor.

Second line treatment includes metronidazole instead of clarithromycin, in which case the patient MUST NOT drink alcohol.

Metronidazole has an antabuse effect; a patient taking metronidazole and alcohol may suffer flushing, vomiting, palpitations, dizziness and chest pain.

The distinction candidate will check the patient has never had eradication treatment in the past. Serology remains positive after treatment, therefore a breath (urease) test or stool antigen test would be needed to prove continued, or recurrent, infection in a patient who continues to have symptoms after eradication therapy.

20 Explanation   Candidate role

You are a Foundation doctor in gastroenterology, and have been observing the colonoscopies today. The patient is 72 years old. S/he has been suffering from episodes of abdominal pain and diarrhoea.

Colonoscopy shows diverticular disease. Your consultant has asked you to explain the diagnosis and treatment to the patient, now that the sedative has worn off.

You do not need to take a history for this station.

20 Explanation   Patient role

You are 72 years old.

You have been suffering from episodes of stomach pain and loose bowels, on and off for about a year now. Your GP referred you to the hospital and you have just had a telescope test, where the doctor looked inside your bowels with a flexible camera. You are feeling a bit more awake now (you had a sedative) and are waiting to hear the results of the test.

Background

You have been quite worried. The GP did tell you that there was a possibility that it might be cancer. You really don't know how you would cope. You live on your own now and like to be independent.

You will be very relieved if it is not cancer... so long as it does not mean an operation. You would not want an operation. You don't want one of those bag things - even though you have heard that Cliff Richard has one!

In fact it has been quite hard with these episodes of pain over the last year. It is very difficult for you if you are ill and you are all by yourself. You hope that the doctor has some treatment that will sort it all out. Yes, that is what you want, some tablets to sort it out.

20 Explanation

Examiner mark sheet

Introduces self and identifies role

Checks patient's identity

Confirms/ establishes reason for visit

Establishes understanding of test and of the expected result

Facts to include in the explanation:

> What the result is
>
> *   The tests shows that there is no sign of cancer
> *   There are some weaknesses in the wall of the large bowel that have caused little out-pouchings, or pockets to form. These pockets can trap stool and cause pain. They are called diverticulae, so the problem is called diverticular disease.
>
> What can be done about it
>
> Treatment options include:
>
> *   A high fibre diet to help keep the bulk moving through the colon
> *   Bulk forming laxatives (such as fybogel or movicol) to help as above
> *   Simple pain killers for mild episodes of pain
> *   Surgery may be needed if the diverticulae repeatedly cause bad pain, or if they cause complications
>
> Safety netting
>
> *   Complications of diverticulae can cause severe pain, fever or vomiting. They should see the doctor straight away as they are likely to need hospital admission for a drip and antibiotics into a vein.

Uses chunking and checking - giving small pieces of information and checking understanding before continuing

Establishes patient's concerns, validates and addresses these (likelihood of cancer; colostomy)

Checks if patient has other questions

Uses communication tools, such as signposting and summarising, appropriately

Offers written advice or websites

Language appropriate throughout

Friendly approach, appropriate body language

Fluent and professional manner

If the patient asks about surgery and colostomy, then explain that a small number of people who have surgery for diverticular disease do need a colostomy bag, but by no means everyone. Sometimes it is possible to join the ends of the cut bowel together either immediately or at a later operation. Just now a colostomy is not anticipated for this patient.

20 Notes

21 Explanation Candidate role

You are a Foundation doctor in General Practice.

Your next patient saw the GP two months ago, complaining of recurrent right upper quadrant pain brought on by eating fatty food. The GP ordered an ultrasound scan. The patient has been asked to attend today for the result. It shows gall stones. There is no other abnormality. Please discuss this with the patient.

21 Explanation   Patient role

You are 35 years old.

You saw your GP a couple of months ago with pain in your abdomen. He sent you for an ultrasound scan and now you have come in for the result.

The pain hasn't been so bad since last time you saw the doctor. He told you to avoid rich meals and junk food, which has helped. The GP last time also gave you some painkillers (co-codamol and diclofenac), but in fact since you cut down on the fatty foods you have only had the pain once.

When the pain was there it just below your ribs on the right and it was quite severe. It lasted for a couple of hours. You tried antacids but they didn't work. You are not sure if the painkillers from the doctor made any difference or not.

Background

You work in a fast food chain which has not made avoiding fatty food easy, so you are quite proud of how well you have done! When you get your breaks you used to eat a burger and chips but you have been sticking to salads (no dressing) the last couple of months and the change in your diet does mean that you have lost a few pounds!

You would not want to take too much time off work as the company is already talking about laying people off.

If the doctor mentions gall stones, you will remember your mother had an op for gall stones when you were a child. You remember that it was quite a big operation. She was in hospital for a week and off work for a couple of months. You don't think you could afford that sort of time off. Maybe just sticking to the new diet would be better?

You are married with two children. You smoke five cigarettes a day and drink a couple of beers twice a week. You are a little overweight but improving now!

21 Explanation   Examiner mark sheet

Introduces self and identifies role
Checks patient's identity
Confirms/ establishes reason for visit
Establishes understanding of test and of the expected result
Facts to include in the explanation:

- What the result shows

- The liver makes bile to help digest (breakdown) fatty food.
- Bile is stored in the gallbladder and squeezed into the gut when we eat fatty food.
- Sometimes stones form in the gall bladder = gall stones.
- Gallstones do not always cause a problem, and if they are found by chance they can be left alone

Why it is important

- They can cause pain if they block the outlet to the gall bladder when it is trying to squeeze the bile out into the gut. And they can irritate the gall duct and cause infection.

What can be done about it

- When pain occurs due to gallstones, avoiding fatty foods may help, but there is a risk that the gallbladder will get infected, which is a more serious illness. Because of this doctors usually recommend surgery to avoid future complications.
- Removal is usually done by keyhole surgery these days, which means patients do not need to be in hospital for more than a day or two (if all goes to plan).

Establishes patient's concerns, validates and addresses these (time off work and long recovery because of operation) - Offer to find out how long off work if patient wishes to know this.

Uses chunking and checking - giving small pieces of information and checking understanding before continuing
Checks if patient has other questions
Uses communication tools, such as signposting and summarising, appropriately
Offers written advice or websites
Language appropriate throughout
Fluent and professional manner

21 Notes

LEARNING POINTS

For most sedentary occupations return to work after a laparoscopic cholecystectomy will be around two weeks if there are no complications. See www.rcseng.ac.uk/patient_information/get-well-soon

Make sure you understand the options in managing gall stones that are causing pain. In addition to laparoscopic cholecystectomy, there are other treatments available in certain circumstances:

Acute cholecystitis (when the patient has severe pain, fever and rigors) may require the patient to have IV antibiotics and be nil by mouth for a few days.

Gallstones trapped in the common bile duct may be removed by ERCP (the patient swallows a tube with flexible camera and a little hook inside it, so that the surgeon can use the hook to fish out the stone)

A few people chose to try a medicine to dissolve the gall stones, but this is only possible for small gallstones and even then is not always successful.

Lithotripsy is not usually suitable for gallstones as the majority of gallstones are soft cholesterol stones – they cannot be smashed by the ultrasonic waves (unlike hard mineral stones as occur in the kidney).

Always remember that the patient has the right to choose not to have surgery even if you feel that would be the best clinical option. If a patient declines to follow your advice you should:

 Make sure they understand the advice (i.e. are competent to make the decision)

 Try to find out why they choose not to take your advice and if there is a problem that you can help them to resolve

 Agree to support them in their decision

 Ensure they are aware that they can change their mind at a later date

 Document your discussion

22 Consent   Candidate role

You are Foundation doctor in gastroenterology. Your next patient needs consenting for endoscopy.

The patient has been suffering from acid reflux and epigastric pain that has not responded well to proton pump inhibitors. There are no red flag symptoms or signs.

You do not have to take a history for this station.

22 Consent   Patient role

You are 31 years old.

You have been suffering from acid reflux and pain in your upper abdomen, for months now. Your GP gave you some tablets that helped the pain a bit, but the acid still kept coming up. He did some blood tests too (looking for some kind of helicopter bug??!!) but he said that the test was normal. So now you have come in for one of those telescope tests, where they make you swallow a camera.

You are not really looking forward to it as you do tend to choke very easily if you get embarrassed or if you don't chew your food properly. Your aunty had it and said she was put to sleep for it. You hope that you can be asleep too as you are worried that it will be very uncomfortable.

22 Consent   **Examiner mark sheet**

Introduces self and identifies role
Checks patient's identity
Confirms/ establishes reason for visit
Establishes understanding of test and of expected result
Facts to include in the explanation:
What will happen and why (before during and after the procedure)

- The camera is in a flexible thin tube (about the thickness of your little finger – or your stethoscope)
- The back of your throat will be sprayed with local anaesthetic to numb it a little, so that swallowing the camera does not hurt – it is like swallowing something you have not chewed properly
- The procedure only takes about twenty minutes and you can usually watch what is being seen on a video screen
- The endoscopist may take some tissue biopsies. You should see your GP in three weeks for the results.
- You must be awake to swallow the tube but can have sedative to make you sleepy if they wish
- If you have a sedative you may have to stay in the ward for a few hours and will need someone to take you home. Also you should not drive, operate heavy machinery or sign legal documents for 24 hours

Minor, common complications
The procedure is quite safe – you may have a slight sore throat afterwards

Rare, serious complications
Other side effects are rare but you must see a doctor if you get abdominal pain, a temperature or breathing problems afterwards.

Establishes patient's concerns, validates and addresses these (anaesthesia and discomfort during procedure)
Uses chunking and checking - giving small pieces of information and checking understanding before continuing
Checks if patient has other questions
Uses communication tools, such as signposting and summarising, appropriately
Language appropriate throughout
Friendly approach, appropriate body language
Fluent and professional manner

22 Notes

Endocrinology practice scenarios

23 History   Candidate role

You are a foundation doctor in General Practice.

Your next patient is aged 32 years who has come complaining of tiredness.

Please take a history.

At the end of the history please tell the examiner your differential diagnoses, most likely first.

23 History   Patient role

You are aged 32 years. You work in Primark. You have come to the doctor today because you are feeling so tired recently – over the last few months. You can't understand why you are so exhausted. You have always worked hard – you have to as a single parent – but you are used to that. Just recently though you have been having to go to bed about the same time as the children – you feel so old!

Background

You have a full time job and two children (aged 7 and 9). Your Mum helps you with the children. You don't hear from your ex any more. You are not bothered for yourself (Good riddance to bad rubbish) but it is tough on the kids. Work has been quite stressful recently because you have a new supervisor who is a bit of a .... But you get by.

You don't feel you are depressed. You still enjoy a laugh with your mates if you get time.

You do get to the gym once a week, though you have noticed that you are a bit less fit than you used to be. You get tired on the treadmill more easily than you used to. You can feel your heart pounding. Still at least you are not putting on weight. If anything you have lost a couple of pounds – must be all the running about that you do after the kids. You have not had time to diet, in fact you eat the kid's left-overs - you know you probably shouldn't.

You haven't really noticed any other problems. Just the tiredness really. And your heart is racing when you are on the treadmill. Oh and you have noticed that your hair is thinning a bit – you are definitely getting old now...

You just want to find out why you are so tired and to feel better.

Your breathing is OK – except if you are running in the gym, but that is more tiredness and heart racing than breathing problems.

You eat like a horse. No tummy pain or problems with your bowels. You are very regular (twice per day)

If you are a woman then say that your periods aren't a problem – you have the contraceptive implant so they are very light.

Your eye sight is OK, though your eyes have been a bit gritty lately. You thought it was just hay fever.

If the doctor asks you, you have been fighting with the children about the heating. They keep turning the heating up and you have to turn it down – they must think you are made of money. You only have to wear a thin Tee shirt in the house so why do they keep complaining of cold?

You are on no medication. You don't smoke (can't afford it) but you drink a couple of glasses a week to relax. You have only been in hospital to have your children.

Your Mum has diabetes takes tablets for it, and your dad has arthritis but otherwise everyone in the family is OK.

23 History   **Examiner mark sheet**

Introduces self and role and checks patient's identity
Explains purpose of interview
Establishes nature of presenting complaint
- time duration
- life changes associated with onset

Enquires about relevant associated symptoms:
Psychological
- mood, enjoyment of activities, sleep
- social stressors

General
- weight change
- appetite
- pain

Thyroid symptoms
- Palpitations – triggers, character, associated pain or breathlessness
- Changes in neck
- Changes in eyes
- Heat / cold intolerance
- Thinned hair

Anaemia
- Bleeding – bowels, urine, periods (if patient female)
- Indigestion

Establishes effect on daily life
Establishes patient's ideas, concerns and expectations
Drug history (especially NSAIDS which can cause GI bleeds)
Establishes previous medical history and allergies
Establishes family medical history
Establishes Social history
- occupation
- who is at home
- smoking
- alcohol

Excludes other systemic symptoms
Appropriate questioning technique (mixture of open and closed questions)
Avoids or explains jargon
Uses tools such as signposting and summarising
Systematic, logical approach
Checks if the patient has any other questions
Friendly approach, appropriate body language
Finishes consultation appropriately

23 Notes

Differential diagnoses
Hyperthyroidism
Stress
Exclude anaemia (and look for cause)
Exclude diabetes

24 History   Candidate role

You are a Foundation doctor in General Practice.
Your next patient is a 43 year old woman, complaining of tiredness.

Please take a history.
At the end of the consultation, please tell the examiner
- your differential diagnosis
- the investigations that you plan.

24 History   Patient role

You are a 43 year old woman who works in IT.

You have been feeling so tired recently that you have decided to come to the doctor to see if there is something wrong. It is all you can manage to keep the house tidy and cook for the kids. The kids have started complaining that you will not go swimming with them at weekends (you let them go by themselves and you stay at home and take a nap). You find you are having to go to bed at 9 o'clock when the kids go, and yet you are still struggling to get up in the morning. You have never liked getting up but really now you could sleep for England. It has been a couple of months now.

You don't think you are depressed. Life has been a bit tough since your husband left you with the kids a couple of years ago but you have managed. You have had an occasional weep with a glass of wine but generally you think this is something physical. You feel like you are getting old before your time.

Maybe you just need a holiday? Somewhere hot away from the horrible cold English weather... nice blue skies might make you feel better

Background – only give this information if asked

You have not noticed any other symptoms. Your periods are a bit heavier than they used to be, but you assumed that was just your age. Your periods started when you were 13 and are regular – every 30 days you bleed for three or four days.

You are eating OK. You have put on a couple of pounds lately, probably because you are too tired to go to the gym any more.

Your bowels are OK. You have no problems with your waterworks.

You have always preferred hot weather (only say if asked).

You could do with finding time to go to the hairdresser but otherwise you would not say that you have noticed any changes in your hair or skin.

There are still some things that you enjoy doing –you enjoy a snuggle up on the sofa with them to watch a DVD on a Saturday night.

You drink maybe a bottle of wine a week and smoke one cigarette a day (the children don't know you smoke).

You are taking herbal tablets that are supposed to give you energy – some kind of Amazonian herb. Otherwise you are not on any medicines.

You were in hospital when you had the children, but otherwise you have not had any operations.

Your mother has high blood pressure and your father has just been diagnosed with angina.

24 History   Examiner mark sheet

Introduces self and role and checks patient's identity
Explains purpose of interview
Establishes nature of presenting complaint
- time duration
- life changes associated with onset

Enquires about relevant associated symptoms:
Psychological
- mood, enjoyment of activities, sleep
- social stressors

General
- weight change
- appetite
- pain

Thyroid symptoms
- Palpitations – triggers, character, associated pain or breathlessness
- Changes in neck
- Changes in eyes
- Heat / cold intolerance
- Thinned hair

Anaemia
- Bleeding – bowels, urine, periods (if patient female)
- Indigestion

Establishes effect on daily life
Establishes patient's ideas, concerns and expectations
Drug history (especially NSAIDS which can cause GI bleeds)
Establishes previous medical history and allergies
Establishes family medical history
Establishes Social history
- occupation
- who is at home
- smoking
- alcohol

Excludes other systemic symptoms
Appropriate questioning technique (mixture of open and closed questions)
Avoids or explains jargon
Uses tools such as signposting and summarising
Systematic, logical approach
Checks if the patient has any other questions
Friendly approach, appropriate body language
Finishes consultation appropriately

24 Notes

Differential diagnosis

Hypothyroidism
Anaemia (due to heavy periods)
Depression
Stress
Diabetes mellitus
Post viral fatigue

Investigations

FBC
Thyroid function test
Glucose
U&Es
LFTs
CRP

25 Explanation   Candidate role

You are a Foundation doctor in General Practice

Your next patient is a 32 year old who has been seeing the GP for investigation of tiredness and weight loss. The patient has come now for the results of the blood tests.

The results show that he is hyperthyroid, with thyroid receptor antibodies present. Please discuss the results and possible treatments.

You do not need to start treatment today – you can discuss his case with your GP trainer and bring the patient back next week. You do not have to take a history for this station.

25 Explanation   Patient role

You are aged 32.

You have been under investigation by your doctor because you have been feeling tired all the time and had noticed that you had lost a few pounds. You had also noticed a slight swelling in your neck which was quite worrisome.

The doctor took blood last week and now has asked you to come in to talk about the results of your recent blood tests

You will be very concerned to know what treatment is going to be available, although you might not start treatment today.

25 Explanation

Examiner mark sheet

Introduces self and identifies role
Checks patient's identity
Confirms/ establishes reason for visit
Establishes understanding of test and of the expected result
Facts to include in the explanation:
What the test shows
- The thyroid gland is overactive

What that means
- thyroid is a gland in the neck that controls metabolism, so over-activity causes weight loss, tiredness, palpitations and can cause swelling in the neck
- explains cause briefly - the body is reacting against itself

What can be done - discusses treatment options in brief:
Tablets (carbimazole)
- to slow the thyroid down,
- taken for several months, with regular blood tests to check on the response
- problem may recur after stopping tablets in which case alternative would be recommended

Radioiodine
- given as a drink in hospital
- the thyroid gland concentrates the iodine so the effect of the radiation is to act only on the thyroid

Surgery is an alternative, especially if there is a large goitre
- scar in necklace line

Mentions that even after treatment patient is likely to need blood tests once per year to make sure the hormone level is correct.
Most often patients need thyroid hormone replacement after treatment as the thyroid may go under-active.

Establishes patient's concerns, validates and addresses these (what the treatment will be and when it will start)
Uses chunking and checking - giving small pieces of information and checking understanding before continuing
Checks if patient has other questions
Uses communication tools, such as signposting and summarising, appropriately
Offers written advice or websites
Language appropriate throughout
Friendly approach, appropriate body language
Fluent and professional manner

25 Notes

Useful website

http://www.btf-thyroid.org/

26 Explanation   Candidate role

You are a Foundation doctor in General Practice.

Your next patient is an obese 45 year old who has been seen on a couple of occasions with abscesses. Your trainer took a fasting blood sugar and the result has come back as 13mmol/l.

Please discuss this result with your patient.

26 Explanation   Patient role

You are 45 years old and obese.

You have seen the GP a couple of times recently with boils on your back, apart from that you have been fine. You know you are overweight but everyone in your family is big.

After you had the second boil treated the GP took a blood test when you had had nothing to eat. He said the infections might be caused by a high blood sugar. When you phoned the surgery for the results you were told to make an appointment, so now you are a bit worried that you might have diabetes. You had an aunty with diabetes and she ended up blind with kidney disease. You would be very anxious about anything like that. And there is no way you could give yourself injections. Isn't that what diabetics do? If the doctor suggests any ways to avoid injections you would be very keen to try these.

26 Explanation   Examiner mark sheet

Introduces self and identifies role

Checks patient's identity

Confirms/ establishes reason for visit

Establishes understanding of test

Facts to include in the explanation:

> What the result shows

- Blood sugar test was high confirming the patient has mild diabetes

> What that means/ why it is a problem

- Diabetes is increasingly common nowadays, probably because our sedentary lifestyle and plentiful food mean our bodies cannot handle the sugar as we get older
- Uncontrolled diabetes can cause problems as high sugar damages different organs in the body, including the eyes, the kidneys, the blood vessels and heart, giving a risk of serious complications in later years

> How it can be treated/ complications avoided

- By controlling the diabetes the risks can be controlled
- Weight loss to get to a healthy weight is one of the most important ways of helping control the diabetes
- Exercise will also help
- Avoid smoking as this is a risk in itself to the heart
- Some people with diabetes need to go onto tablets, and a few will need insulin injections but this is usually only if diet and exercise don't work.

> Follow up

- You will arrange for the patient to see a nurse, health trainer or dietician for tailored dietary advice and further discussion about how to improve the lifestyle.
- The patient will need some more blood tests and a urine test and will get regular check ups to see how they are managing with the lifestyle changes
- You will arrange for the patient to have an eye check every year and a foot test

Uses chunking and checking - giving small pieces of information and checking understanding before continuing

Establishes patient's concerns, validates and addresses these

Checks if patient has other questions

Uses communication tools, such as signposting and summarising, appropriately

Offers written advice or websites

Language appropriate throughout

Friendly approach, appropriate body language

Fluent and professional manner

26 Notes

LEARNING POINT
Reassure the patient that if diabetes is well controlled problems can be
minimised – use it as an educational opportunity.

Neurology practice scenarios

27 History   Candidate Role

You are a Foundation doctor in the General practice out of hours service.
Your next patient is 45 years old.
The patient is complaining of a headache.
Please
- take a relevant history
- then tell the examiner your differential diagnoses
- and the investigations (if any) you plan to carry out.

27 History   Patient role

You are 43 years old.

You have come to see the out of hours doctor because you have a really bad headache (9/10 on the pain scale). You have had headaches in the past, but not like this. You were at home getting the children's clothes ready for school tomorrow. Well, actually you were on the phone to your ex. You had a bit of a row (as usual) and in the middle of the phone call you felt this headache come on. It just came out of the blue, right across the back of your head, like your head was exploding.

The pain is still there, (that is about three or more hours now). You took a couple of nurofen, which is what you usually take if you get a headache, but it has not settled. You feel a bit sick too, maybe the nurofen has upset your stomach a bit. You just can't do anything for the pain. It is not like you.

If the doctor asks whether the light hurts your eyes, respond positively but qualify it with something along the lines of "er, well, come to think of it, it is a bit bright in here. I am usually better in a dark room when I have a bad head."

If the doctor asks about other symptoms, you do not have any numbness, weakness or funny feelings and did not black out with the headache.

Background

Your usual headaches are kind of like a band across the front of your head or round your temples. You get them every now and again, especially if the kids are playing up. But they usually settle if you take the nurofen and have a break.

Before this headache came on, you were feeling fine. No coughs or colds or temperatures, nothing.

You are usually quite well. You have been on antidepressants (citalopram once a day), since your relationship broke up but really you are much better and were thinking of coming off them.

You really just want to get something for this pain and get back home. You had to get a neighbour round to look after the children, and you have to get up to open up your shop in the morning.

27 History   Examiner mark sheet

Introduces self and role and checks identity of patient

Explains/ confirms purpose of interview

Establishes nature of presenting complaint (as relevant)

- site
- onset – sudden, gradual or with warning (aura)
- character
- radiation
- time duration
- exacerbating and relieving factors, (note movement, coughing, sneezing or bending make raised intracranial pressure worse but can exacerbate many headaches)
- severity

Enquires about relevant associated symptoms:

- photophobia
- other visual symptoms such as flashing lights
- nausea/ vomiting
- numbness
- weakness/paresis/paralysis
- speech difficulties
- neck stiffness
- fever
- rash

Establishes effect on daily life

Explores ideas, concerns, expectations (wants to get relief from the pain; go home to open the shop)

Drug history (including over the counter and illicit drugs)

Establishes previous medical history including allergies

Establishes family medical history

Establishes social history

- occupation
- who is at home
- smoking
- alcohol

Excludes other systemic symptoms

Appropriate questioning technique (mixture of open and closed questions)

Avoids or explains jargon

Uses tools such as signposting and summarising

Systematic, logical approach

Checks if the patient has any other questions

Finishes consultation in suitable manner

Friendly approach and appropriate body language

27 Notes

Differential diagnoses

Sub-arachnoid haemorrhage
Tension headache
Migraine (but not the typical unilateral headache with aura)
Meningitis (unlikely as no fever or systemic upset)

Investigations

FBC – to look at white cell count in case of infection
CRP – signs of inflammation
Imaging – CT/MRI
Lumbar puncture

28 History   Candidate role

You are a Foundation doctor in General Practice.

Your next patient is a 48 year old, who has come to see you complaining of 'funny turns'.

Please take a history.

At the end of the consultation tell the examiner your differential diagnoses, explaining your reasoning.

28 History   Patient role

You are 48 years old.

You have come to see the GP today because you have been having funny turns. They started when you woke up in the middle of the night (the night before last) and it felt like the room was spinning. You got up to get a glass of water and immediately felt sick and had to go to the bathroom to throw up. Since then you have been feeling light-headed most of the time. You have had the spinning attacks about four or five times over the last two days, but not been sick again. The only thing that seems to help is lying down until the feeling settles down (it takes about 10 – 20 minutes).

At first you thought you might be having a stroke. As nothing else has happened apart from the spinning attacks and feeling sick, you think it is not likely to be a stroke (you know that strokes cause problems with your arms and legs and speech). However, you are worried that something serious is going on in your brain, but you have no idea what it might be. And you will need a sick note as you can't go into work like this.

Background:

The funny turn is like dizziness, a spinning sensation, and feeling like you want to throw up. It feels like you are drunk. You also feel a bit disconnected from the world, as though you are dreaming your life – it is difficult to describe. You feel quite anxious as well.

The spinning sensation and nausea improves on lying down and worsens on moving your head, especially suddenly. You feel you are slightly deaf – your hearing is normally fine.

You have had a sore throat for about five days, but you are normally fit and well.

You are on no medication.

Your only other medical problem is piles.

You are allergic to penicillin – it brings you out in blotches.

You work in the police force and live alone, since your marriage broke up.

You drink a couple of bottles of wine a week, but this is definitely not alcohol causing this. You have not smoked since your early twenties.

28 History   Examiner mark sheet

Introduces self and role and checks identity of patient
Explains/ confirms purpose of interview
Establishes nature of presenting complaint
- onset
- character (clearly distinguishes between light headedness and true vertigo)
- time duration
- exacerbating and relieving factors
- any precipitating factors

Enquires about relevant associated symptoms:
- vomiting
- deafness
- tinnitus
- ear pain
- ear discharge
- eye problems
- numbness or weakness

Establishes effect on daily life (works in police force)
Explores ideas, concerns, expectations (worried re brain problem; needs a sick note)
Drug history (including over the counter and illicit drugs)
Allergies
Establishes previous medical history
Establishes family medical history
Establishes Social history
- occupation
- who is at home
- smoking
- alcohol

Excludes other systemic symptoms
Appropriate questioning technique (mixture of open and closed questions)
Avoids or explains jargon
Uses tools such as signposting and summarising
Systematic, logical approach
Checks if the patient has any other questions
Finishes consultation in suitable manner
Friendly approach and appropriate body language

28 Notes

Differential diagnosis

Acute viral labyrinthitis
Vestibular neuronitis
Benign positional vertigo
Ménière's disease (no symptoms of tinnitus or deafness so less likely)
Space occupying lesion (e.g. acoustic neuroma)
Multiple sclerosis
Intracranial bleed

29 Explanation   Candidate role

You are a Foundation doctor on a neurology ward.

Your next patient is a 23 year old man who has just been diagnosed with epilepsy. He is to be discharged home on sodium valproate.

Please talk to him about the condition in preparation for his discharge.

29 Explanation   Patient role

You are a 23 year old man.

You had a fit 3 weeks ago and another three days ago. Yesterday you had a scan and lots of other tests in hospital and you have been told that you have epilepsy. You don't remember much about the fits, just waking up in the ambulance feeling really hung over.

You are about to go home but first the doctor is coming to talk to you about your condition and its treatment. You are quite worried about having another fit. Should your family just phone 999? What if it happens at work or in public? It would be so embarrassing.

Isn't there something the doctors can do to cure it? The papers are always full of how much medicine can do these days.

Background:

You live with your Mum and Dad.

You drive to work (in IT for the local council), but it is only a couple of miles and there are regular buses.

In your spare time you enjoy swimming regularly.

29 Explanation   **Examiner mark sheet**

Introduces self and identifies role

Checks patient's identity

Confirms/ establishes reason for talk

Establishes understanding of epilepsy and corrects misunderstandings

Establishes patient's concerns, validates and addresses these (embarrassment from fits)

Facts to include in the explanation:

What epilepsy is:

- repeated fits (seizures) due to a disturbance of the normal electrical activity of the brain
- cause usually unknown, as in this case

What can be done

- treatment aims to reduce the frequency of fits, it does not cure the problem
- Short fits do not cause harm and do not need a hospital admission
- Witnesses should try to keep the patient safe from injury. They can lie the patient on his side. They should not try to put anything between his teeth.
- Tablets must be continued long term, sometimes for life, certainly until advised to come off or change treatment by her doctor
- Some anti-epilepsy tablets reduce the effectiveness of many other medicines, so always let doctors know
- Wearing an alert bracelet or pendant can be helpful

Other issues for epileptics affecting activities of daily life

- Explains DVLA requirements: Patient must inform DVLA and stop driving until epilepsy shown to be controlled for at least a year.
- Discusses work (office work should not be a problem but patient may like to let employers know of diagnosis in case he has a fit at work)
- Discusses need to avoid other activities which would put patient at risk if he had a fit (swimming alone, working at heights or with heavy machinery)

Uses chunking and checking - giving small pieces of information and checking understanding before continuing

Checks if patient has other questions

Uses communication tools, such as signposting and summarising, appropriately

Offers written advice or websites

Language appropriate throughout (avoids or explains jargon)

Fluent and professional manner

Friendly approach, appropriate body language

29 Notes

LEARNING POINT

We made this station easier for you by giving you a male patient. Remember that when you treat women of reproductive age you must always think about how the illness or medication would affect contraception and pregnancy. Epileptic women should be encouraged to plan ahead if they wish to get pregnant to ensure that they are on the least teratogenic medication, well controlled and taking extra folic acid.

30 Explanation

Candidate role

You are a Foundation doctor in General Practice.

Your next patient is a 21 year old woman diagnosed nine years ago with epilepsy. She has come to see you about her medication.

Please talk to her and answer her questions

30 Explanation   **Patient role**

You are a 21 year old woman with epilepsy.

You have had epilepsy since you were ten, and you take epilim (sodium valproate). You have not had a fit now for a three years. Your epilepsy was much worse when you were a child, before the doctors got the tablets right.

You have come to see the doctor today because you are thinking of starting a family. You have been with your boyfriend (Dave) for six months. You have just moved in together and now you are talking about how nice it would be to have a little one. Your sister has just had a baby boy with her boyfriend and is very happy. It has made you broody.

You know that the tablets can cause birth defects in babies. Although you are afraid of your fits coming back, you would never forgive yourself if there was a problem with the baby because of something you had taken. So you want to stop the tablets. You really just came to tell the doctor that...

If the doctor mentions seeing the specialist, you would not be very keen. You remember when you were younger the consultant was very strong about you taking the tablets. Maybe if you see the specialist they won't let you stop.

Background

You do not have any other health problems.

Your periods are regular – about every 28 or 29 days, and last for 3 or 4 days. They started when you were 12.

You are using condoms for contraception.

You are learning to drive. Dave is teaching you.

You do not do any sports.

You don't smoke but you do drink a couple of lagers on a Friday and a Saturday night.

You work on the checkout at the local supermarket.

30 Explanation   Examiner mark sheet

Introduces self and identifies role

Checks patient's identity

Confirms/ establishes reason for visit

Establishes patient's concerns, validates and addresses these (congratulating her on thinking ahead):

Facts to include in the discussion:

- Fitting in pregnancy is also dangerous both for mother and child
- Some epilepsy medication increases the risk of birth defects in children (such as spina bifida and cleft palate). Sodium valproate is one of these.
- Some newer anti-epileptics are thought to be safer in pregnancy, so it is important to see a specialist to see if the medication could be changed
- Need also to weigh up other issues such as driving (which will be prohibited if fits return if medication stopped)

Other pre-pregnancy advice

- Epileptics should take high dose folic acid (5mg) in pregnancy
- Health education for all women contemplating pregnancy on smoking and alcohol

Discusses contraception to be used until specialist advice available

- current use of condoms OK if reliably used but offer post-coital contraception if condoms fail (or more efficacious contraception) until pre-pregnancy care sorted

Uses chunking and checking - giving small pieces of information and checking understanding before continuing

Checks if patient has other questions

Uses communication tools, such as signposting and summarising, appropriately

Offers written advice or websites

Language appropriate throughout (avoids jargon)

Fluent and professional manner

Friendly approach, appropriate body languages

30 Notes

> **LEARNING POINT**
> Students may pass this station even if they are not sure which anti-epileptics are teratogenic, provided they refer to a specialist appropriately, consider other aspects of pre-pregnancy care and manage the consultation well.

Urology practice scenarios

31 History   Candidate role

You are a Foundation doctor in general practice.

Your next patient is a 61 year old man who is complaining of problems in passing urine.

Please take a history from him.

Explain any examinations and investigations you would like to carry out and possible treatment options.

31 History   Patient role

You are a 61 year old man. You have come to see your GP about your problems passing water.

For the last year or so you have been finding it harder and harder to pee. It can take you up to a minute to start and then it takes a long time. Other men can have been and gone in the urinals whilst you are still standing there. It is quite embarrassing, and worse, you have started to dribble a bit at the end. You are very aware that this might make your pants smell, which is not a nice thought. There is no pain and certainly no blood in your urine. Sometimes you have to get up two or three times a night to pee too which disturbs your sleep. Your wife is talking about you having single beds because she is so fed up with you getting up in the night. You are hoping this will not mean the end of your sex life (but you are unlikely to mention this unless you feel really comfortable – and then it may be in a jokey way)

Background

You work as an accountant for Boots the chemist. You do not drink alcohol, but you smoke about 10 – 15 cigarettes per day. You do not take any recreational drugs.

You are otherwise fit and well, though your GP says you have slightly raised blood pressure. She has given you advice about your diet and exercise which you are careful to follow. You are not on any medication and you do not have any allergies.

Your parents have been dead for some time – both died of cancer in their seventies. You remember that your father had problems with his waterworks – his prostrate – and needed an operation. You think you might have the same problem and are expecting that you will need an operation as well. You have one older brother who has a stomach ulcer.

31 History   # Examiner mark sheet

Introduces self and role and checks identity of patient

Explains/confirms purpose of interview

Establishes nature of presenting complaint:

- onset
- time duration
- obstructive symptoms (hesitancy; poor flow; intermittency; terminal dribbling; residual urine)
- irritative symptoms (dysuria; urgency; frequency; nocturia)
- incontinence (stress incontinence; urge incontinence; continuous incontinence)
- renal and ureteric colic (felt in the loin; colicky pain)
- male sexual dysfunction (impotence; retrograde ejaculation)

Establishes effect on daily life (poor sleep, marital upset)

Explores ideas, concerns, expectations (concerns over smell and over separate beds ending sex life; expects an operation)

Drug history (including over the counter and illicit drugs)

Allergies

Establishes previous medical history

Establishes family medical history

Establishes Social history

- occupation
- who is at home
- smoking
- alcohol

Excludes other systemic symptoms

Appropriate questioning technique (mixture of open and closed questions)

Avoids or explains jargon

Uses tools such as signposting and summarising

Systematic, logical approach

Checks if the patient has any other questions

Finishes consultation appropriately

Friendly approach and appropriate body language

31 Notes

Examination
Abdominal examination (to check for a palpable bladder)
Rectal examination (to feel the prostate)

Investigations
MSU
PSA (after counselling the patient about false positives for prostatic cancer)
U&Es

Treatment options depend on findings
For benign prostatic hyperplasia, treatment options include
- alpha-blockers (such as doxazosin) which act quickly to relieve symptoms by relaxing smooth muscle
- 5-alpha-reductase inhibitors (such as finasteride) which act over a period of weeks to reduce growth of the prostate by reducing dihydrotestosterone levels.
- Surgery may be considered if more conservative measures have failed

In this case an alpha blocker may be a first choice of drug because it has a side effect of lowering blood pressure (which may be beneficial), whereas the alpha 1 reductase inhibitor may cause impotence – a concern for this patient.

For prostatic cancer, treatment may include a variety of options according to the stage and grade of the tumour. Treatments include medicines (many of which are anti-androgens), surgery, radiotherapy, high intensity focused ultrasound and cryotherapy.

LEARNING POINTS

In the UK there is no screening programme for prostate cancer at present. This is because the tests used for prostate cancer are not adequately sensitive and specific to detect only those cancers which would have caused the individual harm during his lifetime. At present tests also pick up a large number of small cancers that would never have caused harm during the individual's lifetime, and the treatment of these could cause considerable morbidity and mortality itself.

Before taking blood for a PSA test, you should discuss the pros and cons of the test with the patient so that the patient makes the decision to be tested in full knowledge of the problems that the test incurs.

Look at the website www.cancerscreening.nhs.uk for information about all the cancer screening programmes in the UK. The site also has very good patient information leaflets to inform your discussion about prostate specific antigen testing.

---

32 History   Candidate role

You are a Foundation doctor in urology outpatients.

Your next patient is a 59 year old woman, who has been referred to the clinic complaining of problems with her water.

Please take a history from her and explain any examinations or investigations you would like to carry out.

32 History   Patient role

You are a 59 year old woman. About two months ago you noticed that your water was dark in colour, and it has been like that on and off since. When you visited your GP a couple of weeks ago, he tested a sample and said there was blood and protein in it. Your GP has sent you to see a specialist because of this.

Background

You have not had any other problems passing water, or any other problems that you can think of, certainly no fever, pain or weight loss. You are certain that this is a urinary problem. You have no bleeding from the back passage and no tummy pain. Your periods stopped when you were 51 and you have had no bleeding from the front passage since.

You are worried that you are bleeding, as you read on the internet that it could be a sign of bladder cancer.

You work in grocer's shop.

You like an occasional half of Guinness (maybe at a weekend) and you do not smoke cigarettes. You do not take any recreational drugs and might joke about it if asked.

You get arthritis in your joints – usually your hands or knees, and it seems to vary with the weather more than anything. You take brufen when the arthritis is bad; you do not have any allergies.

The only real illness you have had in the past was TB which you were treated for in your teens. You remember being on medicines for months and having lots of chest X-rays.

You live on your own. You partner died of ovarian cancer ten years ago.

Your father died of a heart attack when he was 65, your mother lives in sheltered housing, she has had a couple of small strokes, but they didn't leave her with any noticeable problems. You had a sister who died of non-Hodgkin's lymphoma a few years ago.

32 History   Examiner mark sheet

Introduces self and role and checks identity of patient

Explains/confirms purpose of interview

Establishes nature of presenting complaint:

- onset
- time duration
- irritative symptoms (dysuria; frequency; urgency; nocturia)
- incontinence (stress incontinence; urge incontinence; continuous incontinence)
- renal and ureteric colic (felt in the loin; colicky pain)

Establishes associated symptoms

- fever;
- tiredness
- weight loss

Establishes effect on daily life

Explores ideas, concerns, expectations (worried about bladder cancer after reading up on it – should be reassuring without offering false guarantees as this remains one of the possibilities)

Drug history (including over the counter and illicit drugs)

Allergies

Establishes previous medical history

Establishes family medical history

Establishes Social history

- occupation
- who is at home
- smoking
- alcohol

Excludes other systemic symptoms

Appropriate questioning technique (mixture of open and closed questions)

Avoids or explains jargon

Uses tools such as signposting and summarising

Systematic, logical approach

Checks if the patient has any other questions

Finishes consultation appropriately

Friendly approach and appropriate body language

32 Notes

Examinations needed

- Abdominal
- Blood pressure

Investigations needed

Urine tests:
- Mid stream urine for microscopy, culture and sensitivities (to rule out infection)
- Urine cytology
- Early morning urine (x3) for TB microscopy and culture (not done routinely as renal TB is uncommon in Britain but do think of it in view of the history of past TB here. It would be relevant also in those from countries with high TB prevalence. Renal TB in fact presents more often with sterile pyuria than with haematuria)

Blood tests:
- FBC
- Clotting
- U&Es

Renal ultrasound
Cystoscopy

> LEARNING POINT
> The most likely diagnosis after an episode of haematuria depends on the age of the patient, their associated symptoms and their personal history. Many areas now have a one stop clinic available for investigation of haematuria. This allows the patient to have all the investigations and get most of the results in the same day.
> Note that this patient has been referred to the urology clinic under the two week wait rule, as she fits the criteria for suspected urological cancer.

33 Consent   Candidate role

You are a Foundation doctor in urology.

Your next patient, a 59 year old woman, is scheduled for a routine flexible cystoscopy. The surgeon wishes to take a tissue sample from the bladder wall during the procedure.

Please consent her and deal with any concerns she may have.

33 Consent   **Patient role**

You are a 59 year old woman.

About two months ago you noticed that your water was dark in colour. When you visited your GP, he tested a sample and said there was blood and protein in it. Your GP has sent you to see a specialist because of this. You have been seen in clinic and now you are due to have a test to look into your bladder. You have been told that the doctor may want to take a piece of tissue from the bladder and that you may have to stay overnight.

You are worried about the procedure being painful and whether you are likely to be incontinent afterwards.

33 Consent   # Examiner mark sheet

Introduces self, identifies role and checks patient's identity

Confirms/establishes reason for visit

Establishes understanding of test and of expected result (looking for cause of blood in urine)

Facts to include in the explanation:

What will happen and why

- Flexible cystoscopies are usually day cases under local anaesthetic
- The patient lies on the couch on their back
- The urethral opening, (the tube from the bladder where the urine comes out) is cleaned and a little cold jelly squirted in to numb it.
- The cystosope, a flexible thin fibre optic tube is then passed into the urethra, so that the doctor can see inside the bladder
- The patient will feel their bladder filling, as water is passed through the tube to give the doctor a clearer view
- Procedure may be a bit uncomfortable but should not be painful
- If the doctor does take a tissue sample, the patient will not feel it being taken, the bladder wall does not have pain nerve endings
- The whole thing takes about ten minutes

Afterwards

- The doctor may tell the patient what was seen, but if tissue was taken to send to the lab it may take a few days for the results
- Usually the patient can go home the same day
- They must have someone to take them home and someone to be with them overnight, just in case they are ill in the night (although this is very rare)

Minor, common complications

- The patient may experience burning on passing water for a few days afterwards. She should drink plenty of fluids to ease this feeling
- Some blood in her urine is usual. She need only see the doctor if this is dark red, or goes on for more than a couple of days.
- There is a small risk of a water infection, so she should see her GP if she has a temperature or pain in your back

Rare, serious complications

- If she is unable to pass water she should see a doctor - the urethra, the tube form the bladder, can be swollen after the procedure (rare)
- Rarely there may be damage to the urethra or bladder during the procedure and the patient may need a catheter for a few days

Elicits concerns (pain/incontinence), validates and addresses these

Uses chunking and checking

Checks for unanswered questions

Uses communication tools (e.g. signposting and summarising, appropriately

Language appropriate throughout

Fluent and professional manner

33 Notes

34 History   Candidate role

You are a Foundation Doctor in General Practice.
Your next patient is a 58year old man.
Please take a history from him.
At the end of the history, tell the examiner
- your diagnosis
- any investigations you would like to carry out
- likely management options

34 History   Patient role

You are a 58 year old man.

You have come to the doctor today with rather an embarrassing problem. You have been trying to get up courage to come to see the doctor for a while.

You are a divorcee. Your wife left you eight years ago, and you have not had another relationship until this last year. Now you have met a very nice woman who you get on with very well. BUT things are not Ok in the bedroom. You have only tried to make love to her a few times but each time although you would like to have sex, your penis has just not been hard enough. It is very embarrassing. Although your new lady friend is very good about it you are mortified. It has come to the point now where you don't want to put yourself in the position of failing again. And that is a shame because you really like your new partner and would like things to work out long term.

If the doctor asks:

You still feel sexual urges and want to make love. You do get some erections in the early morning when you wake from sleep and can masturbate to ejaculation. Your penis is straight when erect, it just is not very firm. You are still shaving as normal every day and everything else seems fine.

You never had any problem with your wife but then in the latter years of your marriage sex was only for high days and holidays anyway. You have had no other sexual partners since your youth.

If pushed you might ask the doctor to prescribe you some Viagra – they seem to be the answer to every man's problems don't they?

Background

You work in IT.

You do not smoke, although you used to until about ten years ago (10 per day).

You drink a couple of beers two or three nights per week and share a bottle of wine with your lady friend at the weekend.

You are not on any medication and don't remember the last time you saw the doctor.

Your father died of a heart attack in his sixties and your mum has vascular dementia and is in a home now.

34 History   **Examiner mark sheet**

Introduces self and role
Checks identity of patient
Establishes nature of presenting complaint
Explores presenting complaint
- Onset/ duration
- Clarifies loss of erection versus loss of libido
- Every time patient has sex or just sometimes?
- Length and state of current relationship
- History in past relationships
- Morning erections?
- Any bend in penis (Peyronie's disease)

Drug history
Allergies
Past medical history – especially history suggestive of cardiovascular disease or diabetes
Family history
Social history
- occupation
- who is at home
- alcohol
- smoking

Appropriate questioning technique
Avoids of explains jargon
Uses tools such as signposting and summarising
Systematic, logical approach
Checks if the patient has any other questions
Finishes consultation appropriately
Friendly approach and appropriate body language

34 Notes

Diagnosis

Impotence – possibly due to performance anxiety
Other possible causes would include cardiovascular disease, undetected diabetes, hypogonadism (unlikely as no other signs of loss of sexual characteristics), neurological problems (again unlikely to present with impotence as the only presenting symptom)

Investigations

Don't forget to examine the patient's testicles and penis (offering a chaperone/ doctor of preferred gender for the examination)
Blood test should include
  • Fasting glucose
  • Fasting lipids
  • TFTs, prolactin, androgens, LH, Sex binding globulin
  • FBC
  • U&E, LFTs, including gamma GT

Treatment options

  • Phosphodiesterase 5 inhibitors (e.g. Viagra like drugs)
  • sublingual apomorphine
  • intracavernosal injections
  • transurethral alprostadil
  • the use of external devices
  • surgery

LEARNING POINT
Treatment of erectile dysfunction is one of the first areas of medical treatment to be openly rationed on the NHS. Only patients with certain medical conditions that give rise to impotence and those suffering from severe psychological distress due to the problem, who have been seen by a specialist, can be prescribed treatment on an NHS prescription. Others have to pay the full cost of the drug. Furthermore guidance indicates that one treatment a week will be appropriate for most patients with erectile dysfunction, although there is some flexibility in this.
See the BNF for details of who can have NHS treatment.

---

35 History   Candidate role

You are the foundation doctor in a walk in centre.

Your next patient is a 25 year old man. He has been triaged by the nurse and is said to be suffering from pain on urination.

Please take a history and advise on the next steps in management of this patient.

35 History   Patient role

You are a 25 year old man.

You have come to the doctor today because you have had pain when you are peeing over the last week. There is no blood and no discharge that you have noticed. You have not had a fever or any other symptoms that you can think of. But it is like pissing glass.

You are normally fit and well. You do not smoke and drink only at weekends – usually vodka shots, maybe eight or nine over a weekend. You do not do drugs.

You work as an accountant and live alone. The pain has made you quite uncomfortable at work and you are realistic enough to know that you need to get it checked out.

Background

You are a gay man with no regular partner. You have had six or seven sexual encounters in the past three months (only when you get lucky!). You both give and receive anal sex and occasionally have oral sex too. You nearly always use a condom.

Your least sexual encounter was at a party ten days ago when you had oral sex with a one night stand. You will be pretty pissed off if you have caught something.

You had an HIV test about a year ago, when you heard that a previous partner was unwell, but thank goodness it was OK. You have not had a full GUM check-up since then.

35 History   **Examiner mark sheet**

Introduces self and role and checks identity of patient
Explains/confirms purpose of interview
Establishes nature of presenting complaint:
- duration
- character of pain
- pain elsewhere?

Associated symptoms
- Urinary frequency
- haematuria
- penile discharge
- fever
- rash
- sore throat
- anal symptoms

Establishes a full sexual history to include:
- number of partners in last three months
- sex of partners
- partners displaying symptoms
- sex vaginal/ anal (giving or receiving) / oral
- barrier methods used

Establishes effect on daily life (uncomfortable at work)
Explores ideas, concerns, expectations (thinks he may have caught an STI)
Drug history (including over the counter and illicit drugs)
Allergies
Establishes previous medical history
Establishes family medical history
Establishes Social history
- occupation
- who is at home
- smoking
- alcohol

Excludes other systemic symptoms
Appropriate questioning technique (mixture of open and closed questions)
Avoids or explains jargon
Uses tools such as signposting and summarising
Systematic, logical approach
Checks if the patient has any other questions
Finishes consultation appropriately
Friendly approach and appropriate body language

35 Notes

Plan

Suggests swabs to check for STI (penile, oral and anal) and offers blood tests to check for hepatitis, syphilis and HIV

36 Consent

Candidate role

You are a Foundation doctor on a surgical team.
Please consent your next patient for a vasectomy.

36 Consent   **Patient role**

You are a 37 year old man.

You have three children, two by your first wife (a girl aged 15, and a boy aged 12 years old) and one by your second wife (a boy aged 1 year old). You have come to the hospital to have a vasectomy as you do not want to have any more children. The doctor is coming to discuss the vasectomy with you.

Background

You are worried about becoming impotent after the operation.

You can make up any other background information if needed.

36 Consent   **Examiner mark sheet**

Introduces self and identifies role

Checks patient's identity

Confirms/establishes reason for visit

Establishes patient's knowledge of vasectomy

Establishes how many children patient has and what the patient's current partner (if any) thinks of the decision

Facts to include in the explanation:

- Discusses irreversibility clearly – e.g. would the patient still want no more children if his current family were all killed in a car crash?
- Discusses failure rate (1/1000)

What will happen during the procedure

- local anaesthetic injection
- small cut in scrotum, above testes
- vas deferens cut and tied
- soluble stitches or tape
- takes about 15 minutes

What will happen after the procedure

- slightly bruised and sore for a few days
- avoid heavy lifting for a week
- two sperm tests will need to be done 2 months after the operation
- continue contraception until the sperm tests are clear

Minor, common complications

- Feeling bruised for a few days
- Risk of bleeding
- Risk of wound infection

Rare complications

- Rarely a small lump may form at the wound site, that may need treatment

Discusses common myths about vasectomy

- Informs patient that sexual function will not be affected and there will be the usual amount of semen – but it will have no sperm in it

Establishes patient's concerns, validates and addresses these (fear of impotence)

Uses chunking and checking - giving small pieces of information and checking understanding before continuing

Checks for unanswered questions

Uses communication tools, such as signposting and summarising, appropriately

Language appropriate throughout

Fluent and professional manner

36 Notes

LEARNING POINT

As with female sterilisation, the partner's consent to vasectomy is not essential but if partners disagree there is more likely to be regret later on, so it is good practice to discuss the partner's views in this consultation.

Vasectomy is a considerably smaller operation than female sterilisation, so statistically is the safer choice for a couple, if all other things are equal.

37 History   Candidate role

You are a Foundation doctor in General Practice.

Your next patient is a 58-year old woman.

Please take a history and, at the end of the consultation, discuss a management plan with the patient.

37 History   **Patient role**

You are a 58 year old woman.

You have come to see the doctor about your incontinence. You have had this for a little while, but have been embarrassed to talk about it.

You wet yourself a little bit most days. For a couple of years now (well possibly more...) you have leaked a bit if you cough or sneeze. This is only a little bit and you have coped by wearing pant-liners and avoiding strenuous exercise which you know makes you leak.

But now it has got a lot worse. For the last couple of weeks you have lost control much more. It happens if you need to go to the loo, and once it starts you leak a lot – you have had to wear really big pads and are still afraid that you will leak in public. It makes you ashamed to go out. You seem to be going to the loo more often too, but when you do there is not always much wee there. There is some pain in the bladder low down sometimes just as you finish weeing.

You really have no idea what can be done about this but you are very keen to get it sorted out, it is so embarrassing.

Background

Your periods stopped about 7 years ago and you have no other problems down there that you know of. You have not had any blood in your water at all.

You have had high blood pressure for a few years and have been on tablets for that for a long time. You saw a new doctor last month and he changed your medicine but you are not sure of the name of the new tablets he gave you. But you don't really think of blood pressure as a problem – the doctor always said not to worry.

The only other thing you have ever had was your appendix out when you were in your twenties.

You take a vitamin tablet every day too because it is supposed to be good for you.

You drink an occasional glass of wine or sherry and you do not smoke. You are a retired shopkeeper.

You live alone – your husband died in a car accident 12 years ago and your two daughters both live in a nearby town. Your parents both died of cancer (bowel and bone) in their sixties. You do not know of any relatives who had the same problem with incontinence.

Your daughters were both born normally with no particular problems. The only other pregnancy you had ended in a miscarriage at about 8 weeks.

37 History   **Examiner mark sheet**

Introduces self and role and checks identity of patient
Explains/confirms purpose of interview
Establishes nature of presenting complaint:
- patient description of problem
- onset and duration
- aggravating/relieving factors
- previous occurrence of problem/ what she has done about it

Associated symptoms:
- urgency
- frequency
- dysuria/ haematuria/ nocturia
- abdominal pain
- feeling of something coming down

Obs and gynae history – number and types of deliveries, timing of menopause/ post-menopausal bleeding
Establishes effect on daily life (embarrassment, avoiding certain situations, using pads)
Explores ideas, concerns, expectations (wants something done)
Drug history (including over the counter and illicit drugs)
Allergies
Establishes previous medical history
Establishes family medical history
Establishes social history:
- occupation/who is at home/smoking/ alcohol

Excludes other systemic symptoms
Management plan should include
- examine the patient (to include an abdominal and vaginal examination)
- urine dipstick +/- MSU
- U&E, Glucose
- Review the medication – perhaps the new tablets are responsible for the worsening of the symptoms
- Referral to incontinence nursing service for full assessment +/- education on pelvic floor exercises
- If conservative measures fail then consider surgery

Appropriate questioning technique (mixture of open and closed questions)
Avoids or explains jargon
Uses tools such as signposting and summarising
Systematic, logical approach
Checks if the patient has any other questions
Finishes consultation appropriately
Friendly approach and appropriate body language

37 Notes

> LEARNING POINTS
> This patient presents with a history typical of stress incontinence but then with more recent urge incontinence to confuse the picture. In such a situation you should always try to think about what has changed recently. In this case it may be the new tablets, or perhaps the patient has recently developed a urinary tract infection?

Urinary incontinence in women may relate to:

Cause	Symptoms	Treatment
Weak pelvic floor	Stress incontinence (leaking small amounts when coughing or sneezing)	Pelvic floor exercises Consider oestrogens for post-menopausal women Surgery or teflon injection to lift the bladder neck
Hyperactivity of the bladder	Urgency (leaking large amounts when needing to pee and needing to go all the time)	Anticholinergic medication Bladder training (delaying peeing deliberately)
Neurological problems	Urinary retention with overflow incontinence causing dribbling	Treat the neurological condition. Urinary problems may be managed by self-catheterisation
Urinary tract infection	Dysuria and worsening of other symptoms	Antibiotics Consider oestrogens for post-menopausal women with recurrent infections
Diabetes mellitus Diuretics Alpha blockers	All may worsen pre-existing problems	Manage diabetes Review medication

Haematology practice scenarios

38 History and explanation   Candidate role

You are a foundation doctor in general practice.

Your next patient is a 58 year old man who had some blood tests last week because of feeling tired all the time. The initial results showed a mild anaemia with a low MCV. Now you have the rest of the results. They show a very low ferritin. B12 and folate are normal.

Please talk to the patient about his results, take a relevant history and discuss what you plan to do next.

38 History and explanation   Candidate role

You are a 58 year old man. You have been feeling very tired recently and came to the doctor for a check-up. Your friend has just been diagnosed with an under-active thyroid gland and you wondered if you have the same. The doctor sent you to the nurse for a blood test. When you came for those results they said you were slightly anaemic and you had another blood test. Now you have been told to make an appointment with the doctor to discuss the results.

Background

You are normally quite well. You have not been in hospital apart from when you had your gall bladder out several years ago. Oh and you broke your leg when you were younger (you came off a friend's motor bike).

You never have any problems with your stomach or with your bowels – well apart from the piles which give you a bit of jip occasionally. They do bleed a bit when they are playing up – you notice blood on the paper but this has been like that on and off for years. Otherwise your bowels are fine – unless you have been taking co-codamol. You take it occasionally for an ache in your leg (the old fracture) and it does bung you up. In fact you used to take nurofen for the leg, but that gave you indigestion. If it is not one end it is the other!

Otherwise your health is fine. You are eating OK, not losing weight (just the opposite in fact), and you are on no medication from the doctor.

You think you eat a reasonable diet – OK you do have a take away at weekends but everyone needs a treat now and then. You always like a Sunday roast.

You don't smoke and only drink a couple of glasses of wine per week.

You live with your better half and teenage daughter, Jenny. You work in a hardware shop.

You don't understand why you had to have more tests for this anaemia. Why they did not just give you iron tablets – isn't that the usual treatment for anaemia?

All these questions sound a bit worrying – you hope it is nothing serious.... If the doctor wants to do more tests you will be quite alarmed. Does the doctor think this is something bad then? Could it be cancer? Your father died of lung cancer.

38 History and explanation　　Examiner mark sheet

Introduces self and identifies role

Checks patient's identity

Confirms/ establishes reason for visit

Establishes understanding of test and of the expected result

Discusses relevant main points of what has happened

- The anaemia is because he is a bit short of iron
- You need to ask a few more questions to find out why he is short of iron
    - o Any bleeding pr? – when does this happen (on paper, in the toilet pan, mixed with stool or after it); how long has it been happening?; Any change in bowel habit? (looser than normal or more constipated?)
    - o Appetite?
    - o Weight loss?
    - o Any indigestion?
    - o Does he take any tablets, especially painkillers (NSAIDs)?
    - o Does he eat much red meat?

Explain the anaemia may be due to bleeding into the stomach - a side effect of nurofen, but that you need to be sure that is all it is, so you would like the patient to have some more tests

Explain that you will refer the patient for investigation of the bowel. You may wish to start with endoscopy or with colonoscopy.

The patient should be seen within two weeks.

Uses chunking and checking - giving small pieces of information and checking understanding before continuing

Establishes patient's concerns, validates and addresses these:

Deal with patient concerns factually and in a reassuring manner but without giving false promises (there is a small possibility this could be a sign of something serious such as a bowel cancer, but it is more likely to be simply bleeding into the stomach because of the tablets. The investigations will help to find out quickly).

Checks if patient has other questions

Uses communication tools, such as signposting and summarising, appropriately

Offers written advice or websites

Language appropriate throughout

Friendly approach, appropriate body language

Fluent and professional manner

38 Notes

> **LEARNING POINT**
> Remember iron deficiency anaemia is not a diagnosis in itself. You need to look for the cause. Think of blood loss from any source (upper GI, lower GI, per urethra, per vagina in women); drugs that may cause bleeding (NSAIDs, aspirin, anticoagulants); and diet. In the developing you world also consider gut parasites (hookworm). Unexplained anaemia in older people should prompt an urgent search for a malignancy.

39 Explanation and history   Candidate role

You are a Foundation doctor in General Practice. Your next patient is a 29 year old man who had some bloods taken a couple of weeks ago for a health check-up. The blood tests have all come back as normal, other than the following:

Hb   11.2 (ref 12.5 – 14.5)

MCV low

Microcytic hypochromic cells

Ferritin 10 (ref 12 – 200)

Please discuss these results with the patient.

At the end of the consultation the examiner will ask you your diagnosis and management plan if these have not already been discussed.

39 Explanation and history   Patient role

You are a 29 year old man. You attended the practice a couple of weeks to get some blood tests done for a health check-up. When you called the surgery for the results you were asked to make an appointment with the doctor. That sounds worrying…. Now you have come back for the results.

Background

You work as a plumber. Times have been quite tough recently and you and your wife are thinking about emigrating to Australia. You have a job offer over there. In fact the health check was part of the preparations for the big move. The company sponsoring you insist all the boxes are ticked before you go out there.

You are married but no children yet. You don't smoke (too expensive). You drink a couple of beers a couple of nights a week, but otherwise you don't go out much. You are trying to save up for the big trip.

You eat a normal diet – the usual, some junk food, some home cooking, lasagne, or chilli if your wife is cooking, burgers and sausage if you are cooking, curries and pizzas if it is a take away.

You don't normally see the doctor much at all and certainly don't think of yourself as ill. You have never been in hospital. The last time you came to the doctor must have been when you were a teenager, and even then it was football injuries. You don't take any medicines from the doctor.

If the doctor asks you might admit (somewhat reluctantly) the you have been taking nurofen for back pain. It has been giving you a bit of gip on and off for six months now. You think you damaged it when you were helping a friend lifting stuff to build a garden shed. It is just down at the bottom of your back, nowhere else. You have not bothered the doctor about it. You just take the painkillers and keep going as you can't afford time off work.

And you have had to start taking antacid tablets for indigestion in the last couple of months. You get terrible acid coming up behind your breast bone. In fact you keep some Rennies or Settlers Tums by your bed, in your glove compartment in the car, in your locker at work…

You don't really think of yourself as unwell. These are just niggles that you live with. There are certainly nothing that would stop you emigrating. Are they?

Anyway you have no other problems to think of: Your bowels are fine. No problems there. (You don't really look at the colour or to see if there is any blood). Your waterworks have never been a problem.

You want to know if there is a problem and might be a bit worried about something wrong. Is it serious? Is it something that will stop you going to Australia? Or can it be treated and sorted out quick?

39 Explanation and history   Examiner mark sheet

Introduces self and identifies role
Checks patient's identity
Confirms/ establishes reason for visit
Establishes understanding of test and of the expected result
Facts to include in the explanation:

- Test shows the patient is anaemic
- The anaemia is because he is a bit short of iron
- You need to ask a few more questions to find out why he is short of iron
    - o Any bleeding – When he moves his bowels? In his urine?
    - o Any indigestion?
    - o Does he take any tablets, especially painkillers (NSAIDs)?
    - o Does he eat much red meat?
- Explain the likely cause of the anaemia is bleeding into the stomach - a side effect of nurofen.
- Management must include:
    - o stopping the NSAIDs (giving an alternative painkiller, i.e. paracetamol or co-codamol)
    - o treatment of the indigestion (with a proton pump inhibitor)
    - o iron tablets to help the body replace the blood it has lost
- recheck the blood in a couple of months
- Candidates may suggest investigation of the indigestion, including tests for H pylori and / or endoscopy, or may prefer to treat and observe response to treatment in a young man.
- Dietary sources of iron include red meat, dark green leafy vegetables, dried apricots and raisins, some cereals are fortified but it is harder to absorb the iron from these.

Uses chunking & checking - gives small bits of information & checks understanding before continuing
Establishes patient's concerns, validates and addresses these – patient concerned about travel plans; getting this sorted will take a few months but hopefully will not stop the move in the long term.
Checks if patient has other questions
Uses communication tools, such as signposting and summarising, appropriately
Offers written advice or websites
Language appropriate throughout
Friendly approach, appropriate body language
Fluent and professional manner

39 Notes

Differential diagnosis

Iron deficiency anaemia caused by GI blood loss secondary to NSAIDs
More serious causes of blood loss (such as a gastrointestinal cancer are less likely in view of the patient's age.

Management plan

Stop NSAIDs, start PPI, start iron supplements, recheck blood in couple of months to check for response and to review abdominal pain.

> LEARNING POINT
> Think about how you decide what the most likely diagnosis is.
> Compare the suggested management plan for this patient, with that for an older man with a similar problem. This is a good example of how the age of the patient affects the relative probability of the different differential diagnoses. Young people are far less likely to suffer from gastrointestinal cancer than older people, so your investigations should reflect this and you might choose a more conservative (less interventional) approach at first. Of course you would still need to look further if the patient failed to respond to your treatment, so your follow up safety net is important.
> When you give your differential diagnosis list, always try to give the most likely diagnosis first but also mention any dangerous diagnoses you want to rule out.

40 Explanation and history   Candidate role

You are a Foundation doctor in General Practice.

Your next patient is a 21 year old woman who saw the GP a couple of weeks ago complaining of feeling tired. The GP examined her and noted that she was pale. He took a few blood tests, which have all come back as normal, other than the following:

Hb   9.0 (ref 11.5 – 13.5)

  MCV low

  Microcytic hypochromic cells

Ferritin 11 (ref 12 – 200)

Please discuss this with the patient.

At the end of the consultation the examiner will ask you your diagnosis and management plan if these have not already been discussed.

40 Explanation and history — Patient role

You are a 21 year old woman. You attended your GP a couple of weeks ago because you have been feeling tired all the time. The doctor did a blood test and now you have come back for the results.

You had wondered if you were anaemic – your Mum kept on telling you how pale you looked recently and that can be a sign can't it? If the doctor says you are anaemic you will probably be relieved, at least it is nothing worse. You had a friend who died of leukaemia when you were a teenager, so that was at the back of your mind.

But you are not so keen on tablets, is there a more natural way of treating it? You prefer to do things naturally.

Background

You are not on any medicines though you take an infusion of natural oils from the herbal shop to help you sleep. Your sleep is just a bit messed up because sometimes you work night shifts. You work in a children's home.

You used to need inhalers as a child but your chest has improved a lot. You think it might be since you started doing yoga and natural breathing exercises.

Your periods have been very heavy over the last year. In fact you had your coil taken out last month because you think that was the cause. So far so good, this month your period was much lighter.

You have no other health problems, no indigestion, no bleeding, nothing.

40 Explanation and history   Examiner mark sheet

Introduces self and identifies role

Checks patient's identity

Confirms/ establishes reason for visit

Establishes understanding of test and of the expected result

Facts to include in the explanation:

- Test shows the patient is anaemic, the blood is a bit thin which is why she has been tired
- The anaemia is because she is a bit short of iron
- You need to ask a few more questions to find out why she is short of iron
    - Any bleeding – When she moves her bowels? In her urine?
    - Any indigestion? Does she take any tablets, especially painkillers (NSAIDs)?
    - How are her periods?
    - Does she eat much red meat?
- Explain anaemia probably due to heavy periods as a side effect of the coil
- Suggest iron tablets for several months to treat the anaemia and fill up her body's iron stores
- Suggest a recheck of her blood in a couple of months
- Dietary sources of iron include red meat, dark green leafy vegetables, dried apricots and raisins, some cereals are fortified but it is harder to absorb the iron from these.

Uses chunking and checking - giving small pieces of information and checking understanding before continuing

Establishes patient's concerns, validates and addresses these (leukaemia)

Checks if patient has other questions

Uses communication tools, such as signposting and summarising, appropriately

Offers written advice or websites

Language appropriate throughout

Friendly approach, appropriate body language

Fluent and professional manner

40 Notes

Diagnosis

Iron deficiency anaemia due to heavy periods as a side effect of the IUCD

Management

IUCD now removed and periods lighter. Treat iron deficiency with supplements

---

LEARNING POINT

Some medicines need more discussion than others.

Well done if you found the cause of this patient's anaemia and treated her correctly. For the class gold medal, think about how you explained the iron tablets.

A lot of patients start iron but do not take enough to fully treat their iron deficiency. This is because doctors often forget to explain it properly. Remember to warn the patient about potential side effects, and to be clear about how long to take it. Iron can cause black stools, which could be quite alarming for the patient if you have not warned them. They can also cause GI upset (either loose stools or constipation). Sometimes it is worth advising the patient that if the tablets upset them too much when taken three times a day, they can try twice or just once a day but they will need treatment for longer. Treatment of iron deficiency anaemia should continue for several months, so advise the patient of this.

---

41 Explanation   Candidate role

You are a Foundation doctor working as part of a medical team.

Your patient is 32 years old. S/he has been diagnosed with a deep vein thrombosis in the right calf and this diagnosis has been explained to her. S/he has been started on anticoagulation and is now ready to go home on warfarin tablets.

Please give appropriate advice and deal with the patient's concerns.

Warfarin protocol:
- Your patient will go home with warfarin at a dose of 5mg daily.
- S/he will attend the nurse led anticoagulation clinic at the hospital for a blood tests in five days.
- The warfarin dose will be adjusted according to her INR.
- S/he will need continued follow up at the anticoagulation clinic to make sure her/his warfarin dose is correct.
- S/he should stay on warfarin for three months.

41 Explanation   **Patient role**

You are 32 years old.

Your right leg has been painful and swollen over the last couple of days. You had a blood test and a scan and the doctors have told you that you have a clot.

You have had injections and tablets to thin your blood and been shown leg exercises by the physiotherapist. You have been told that you can go home now with tablets, after the doctor has had another word with you.

Background

You are quite worried about the tablets that you have been prescribed to thin your blood. Aren't they rat poison and might they cause you to be ill?

You also want to know what might have happened if the clot had gone to your lungs. You have heard that can be very dangerous. You have mild asthma and take an asthma inhaler. Will that affect the chances of you getting a clot on the lungs?

You are an assistant manager in a Bank and are married with two school-age sons.

You are a busy working and looking after the children so you will not want to have to attend hospital too often if the doctor suggests this - it is difficult to get time off work.

41 Explanation   **Examiner mark sheet**

Introduces self and identifies role

Checks patient's identity

Establishes patient's understanding of the problem (clot in leg which might go to lungs and this would be dangerous)

Establishes understanding of medication

Facts to include in the explanation:

What the medicine is

- Warfarin is a tablet to thin the blood to reduce the risk of another clot

Why it is needed

A clot in the leg is painful but the real problem is that if it breaks off and goes to the lungs it can be life-threatening.

Side effects and risks of medication

- The dose of warfarin varies from patient to patient and has to be checked very carefully. Too much can cause bleeding , too little means another clot may form, so close monitoring is needed
- Patient will need to carry a warfarin card and show it to any health care workers they see, especially if they start other medicines which can interfere with warfare levels
- The patient should see the GP straight away if they get bleeding or bruising

Monitoring

- Blood will be checked in five days at anticoagulation clinic
- Warfarin dose will change depending on the blood results
- Blood will be rechecked every week or so at first, to see how much warfarin is needed, but, once the patient is on the right dose, the frequency of checks will be much less
- Warfarin will need to be continued for three months

For women

- Warfarin can be dangerous in pregnancy, so female patients will need good contraception
- A woman who has had a DVT should avoid the pill in future as it makes clots more likely.

Establishes patient's concerns, validates and addresses these (warfarin as rat poison; clot in lungs; time off for tests)

Checks understanding

Uses signposting and summarising appropriately

Language appropriate throughout

Friendly approach, appropriate body language

Fluent and professional manner

41 Notes

LEARNING POINT

Warfarin like drugs are used as rat poison, this patient is correct. Rats are given a high dose and bleed internally. You can use the knowledge of this to stress to the patient the need to take the right dose and to attend regularly for blood tests.

We have not included here any mention of low molecular weight heparin injections as we felt that this made the explanation too complicated for this station. In reality, patients starting on warfarin are usually started on a low molecular weight heparin injection at the same time. The heparin acts immediately, whereas warfarin takes a few days to get the dose right. The injections can be given by the district nurse at home if necessary, so many patients with a new DVT avoid a hospital admission. Injections are stopped once the warfarin is working.

Can you think of times when continued use of low molecular weight heparin injections may be preferable to starting a patient on warfarin? Hint – low molecular weight heparin injections are not teratogenic, nor do they need frequent blood tests for monitoring.

Note that the newest anticoagulant, dagabatrin, is a tablet that does not require regular blood tests, so it may alter the way we manage anticoagulation in the future.

42 Explanation

Candidate role

You are a foundation doctor in General Practice.

Your next patient is a 62 year old who had some blood tests last week because of feeling tired all the time. The initial results showed a mild anaemia with a high MCV (115 fL). Now you have the rest of the results. They show a very low B12 and serum anti-intrinsic factor antibody is present. Ferritin and folate are normal.

Please talk to your patient about the results and discuss treatment.

42 Explanation   **Patient role**

You are 62 years old. You have been feeling very tired recently and came to the doctor for a check up. Your friend has just been diagnosed with an under-active thyroid gland and you wondered if you have the same. The doctor sent you to the nurse for a blood test. When you came for those results they said you were slightly anaemic and you had another blood test.

Now you have been told to make an appointment with the doctor to discuss the results.

Background

You have had no other symptoms. You do not have any other illnesses (that you know of) and you are not on any medication.

You eat healthily – meat and two veg. for the main meal every day, a bacon sandwich for breakfast and maybe some soup or a salad for lunch. You like cake. You do not drink alcohol or smoke cigarettes.

It all sounds a bit worrying – you hope it is nothing serious....

Why they did not just give you iron tablets – isn't that the usual treatment for anaemia?

You would not like the idea of injections if they were offered – maybe there is some food that you could eat to get enough vitamins?

42 Explanation   Examiner Mark sheet

Introduces self and identifies role

Checks patient's identity

Confirms/ establishes reason for visit

Establishes understanding of test and of the expected result

Facts to include in the explanation:

- Tests show that the patient is short of vitamin B12 (and this has caused anaemia)
- B12 is found in meat, eggs and dairy products but some people can't absorb it properly and this is the case here
- Treatment is injections of B12
- Start with a number of injections to load up the body (e.g. injections twice weekly for four weeks) then after that an injection every three months for life
- Blood can be checked in a few weeks to be sure that patient is responding to injections

Establishes patient's concerns, validates and addresses these (sign of something serious; iron tablets vs. injections)

Uses chunking and checking - giving small pieces of information and checking understanding before continuing

Checks if patient has other questions

Uses communication tools, such as signposting and summarising, appropriately

Offers written advice or websites

Language appropriate throughout (avoids or explains jargon)

Fluent and professional manner

Friendly approach with appropriate body language

42 Notes

> LEARNING POINT
>
> Causes of macrocytic anaemia include B12 deficiency, folate deficiency, alcohol abuse, liver disease and thyroid disease, so remember to check for these if an anaemic patient has a high MCV.
>
> A very low B12 and raised anti-parietal cell antibodies or anti-intrinsic factor antibodies distinguish pernicious anaemia (which is autoimmune) from dietary deficiency.
>
> Vitamin B12 deficiency due to poor diet (e.g. vegans) can be treated with oral B12, and some argue that, given in high enough doses, oral treatment is also adequate for pernicious anaemia but current guidelines still favour 3 monthly intramuscular injections.

Musculoskeletal practice scenarios

43 History   Candidate role

You are a Foundation doctor in General Practice. Your next patient is a 57 year old man, complaining of back pain.
Please:

- take a history,
- tell the examiner what examination you would like to carry out
- and discuss a management plan.

43 History   Patient role

You are a 57 year old man. You have come to see the doctor because you have a pain in your back. You are very worried that you might have seriously damaged your back, because the pain is so bad.

The pain is right down at the base of your spine, just below the level of your hips. It started yesterday after a coughing fit (a cup of tea you were drinking went down the wrong way). The pain is a bad ache (you would rate it 7 out of 10 if asked). The pain was really bad last night and you hardly had any sleep - the paracetamol that you took did not help.

You have had this back ache for several years now. It isn't usually bad, and it comes on after you've been doing some heavy gardening, such as digging. It always got better with rest and paracetamol. The pain you have today is much, much worse than your usual back ache.

You don't have the pain anywhere else, but you started to have some pain and a tingling feeling down the back of your right leg this morning. Your right foot feels a bit floppy – almost like when you have been sitting on your foot and you first stand up. It feels odd to walk on it, like it is sticking to the ground.

Background

Only if the doctor asks you directly would you talk about going to the toilet...Last night and this morning, you found it hard to start peeing and it was like you just passed a dribble. In fact you are a bit embarrassed because you wet your pants with a little bit of a dribble this morning then found you could not really go properly when you went to the loo.

You have not had any injuries recently, or a fever.

Your mood is OK.

You work in a bank as a financial advisor. You are reasonably happy at work. You are normally fit and well, except for a bit of back ache now and again. You do not have any medical problems. You had your tonsils out as a child and you had mumps when you were small. You used to have migraines, but you haven't had a bad one for at least ten years. You do not take any medication except the paracetamol for the occasional backache.

You do not smoke or use recreational drugs. You share a couple of bottles of wine with your wife each week and like to have a glass of beer or whisky when you are on holiday.

43 History   **Examiner mark sheet**

Introduces self and role and checks identity of patient
Explains/confirms purpose of interview
Establishes nature of presenting complaint (as relevant)
- site
- onset
- character
- radiation (right leg)
- time duration
- exacerbating and relieving factors (paracetamol normally, not working now)
- severity
- Has the patient had this problem before?

Enquires about relevant associated symptoms
- Pain in legs, hips, knees?
- Any problems with walking?
- Change in bladder / bowel habit?
- Numbness
- Fever?
- Mood?

Establishes effect on daily life
Explores ideas, concerns, expectations
Drug history (including over the counter and illicit drugs)
Allergies
Establishes previous medical history
Establishes family medical history
Establishes Social history
- occupation
- who is at home
- smoking
- alcohol
- Problems with claims for compensation or applications for benefits?

Excludes other systemic symptoms
Appropriate questioning technique (mixture of open and closed questions)
Avoids or explains jargon
Uses tools such as signposting and summarising
Systematic, logical approach
Checks if the patient has any other questions
Finishes consultation appropriately
Friendly approach and appropriate body language

43 Notes

Examination should include:

Observe patient coming into room
Examine spine
- standing
- forward flexion
- lateral flexion
- rotation

Lying on couch do straight leg raising
Knee and ankle reflexes
Dorsiflexion of foot
Sensation - of saddle area and of right leg/foot
Palpate abdomen to check for a full bladder as the patient reports problems when passing urine

Future investigations and management:

Urgent referral to neurosurgeon (today) for scanning
FBC (white count for signs of infection)
CRP/ESR (infection/inflammation)
U&Es (patient reports urinary problems)
LFTs (raised if cancer with liver metastases)

Differential diagnosis

Possible cauda equina syndrome due to
- prolapsed intervertebral disc
- spinal stenosis
- cancer
- spinal fracture

LEARNING POINT
It is important to be aware of red flag signs in back pain, both in the history and on examination:
-paralysis of limbs below the level of compression
-decreased sensation below the level of compression
-urinary and faecal incontinence and/or urinary retention
-hyperreflexia
These are signs of spinal cord compression. This is a surgical emergency and requires a surgical opinion the same day to prevent nerve damage becoming permanent.

---

44 History   Candidate role

You are a Foundation doctor in General Practice.

Your next patient is a 27 year old. S/he has come to you complaining of back pain.

- Please take a history and examine the patient (or ask the examiner for the clinical findings)
- At the end of the history and examination, discuss with the examiner the likely diagnosis and your management plan.

44 History   Patient role

You are 27 years old.

You have come to see the doctor because you have a pain in your back. You are really fed up with it. It has been a problem for about three months now, aching at the bottom of your back if you try to lift anything heavy.

You want to know what is going on and are sick of being fobbed off with painkillers. You have been in to see the doctors in this surgery a few times now and have not even had an X-ray. It is like they don't believe you. They just say "keep moving" and tell you to take the tablets. You don't think it is right to mask it with tablets all the time, isn't the pain your body's way of telling you something is wrong? You want to know what it is. A scan would tell that.

You were laid off from your job when the shop you worked for went bust and have been signing on for Job Seekers for the last nine months. Now the job centre are on at you about getting a job, but how can you work with this back?

The pain is right down at the base of your spine, just below the level of your hips. It aches most of the time, especially if you try to lift something (you would rate it 7 out of 10 if asked).

You don't have the pain anywhere else.

Only when asked:

You have no numbness or weakness. You are going to the toilet (peeing and moving your bowels) OK, although the painkillers you got last time did bung you up a bit.

You have not had any injuries recently, or other illnesses.

Background

You had your appendix out as a teenager.

You do not take any medication except the painkillers – and you have tried most of those, paracetamol is no use, co-codamol bungs you up and ibuprofen gives you indigestion.

You live with your partner and your partner's nine year old son.

Your Mum has some joint problems – her knees give her a lot of pain. Your Dad has high blood pressure.

You smoke just one or two cigarettes a week, you can't afford more. You drink a few lagers at the weekend. You have found that a joint at the weekend helps the pain but you might not tell the doctor that.

Note

In this station, the patient can allow examination and act someone with a stiff lower back, or simply tell the candidate the examination findings as below.

44 History   Examiner mark sheet

Introduces self and role and checks identity of patient
Explains / confirms purpose of interview
Establishes nature of presenting complaint
- site
- onset
- character
- radiation
- time duration
- exacerbating and relieving factors
- severity

Enquires about relevant associated symptoms:
- Numbness
- Weakness
- Bladder control
- Bowel control

Establishes effect on daily life
Explores ideas, concerns, expectations (wants to know what it is and thinks a scan might help)
Drug history (including over the counter and illicit drugs)
Allergies
Establishes previous medical history
Establishes family medical history
Establishes social history, including occupation / who is at home / smoking / alcohol
Excludes other systemic symptoms
Examination to include:
Examine patient's back from behind (standing) – findings = loss of lumbar lordosis and tenderness over paravertebral muscles
- Ask patient to bend forwards, sideways, and rotate spine – findings as above
- Sensation in perianal area (warn patient and advise why you are testing!)= normal
- Straight leg raising (with patient lying on back) – both legs OK to about 70 degrees
- Knee and ankle reflexes = normal

Appropriate questioning technique (mixture of open and closed questions)
Avoids or explains jargon
Uses tools such as signposting and summarising
Systematic, logical approach
Checks if the patient has any other questions
Finishes consultation appropriately
Friendly approach and appropriate body language

44 Notes

Likely diagnosis - Musculoskeletal back pain

Management Plan:

- Explain to patient that tests such as scans and X-rays are not indicated in this case as they will not change the management of the patient
- Explain that most cases of musculoskeletal back pain resolve in a few months
- Physiotherapy may help by helping the patient to strengthen the muscles of the back and to prevent future problems
- Discuss managing the pain relief – for example give lactulose with co-codamol, or a proton pump inhibitor with a non-steroidal anti-inflammatory drug

---

LEARNING POINT

Managing consultations in which the patient's expectations (for a scan and a cure) are in conflict with good medical practice can be difficult. Lots of explanation is needed and you may still be left with an unhappy patient. Studies have shown that the best indicator of whether a patient with back pain will recover or go on to have chronic on-going problems is actually whether or not the patient believes they will get better – attitude of mind is very important, so you should really try to encourage a positive approach. If you want to really test your communication skills, try this station again but talk to the patient not the examiner about your findings, the diagnosis and your management plan.

45 History/Explanation   Candidate role

You are a Foundation doctor in General Practice. Your next patient is a 35 year old woman, who is complaining of a painful hand.

Please take a history of her complaint, with a view to making a provisional diagnosis.

- You may examine her hand to give you more information.
- When you have reached a conclusion, please discuss with the patient what you think should happen next and why.

45 History/ Explanation   **Patient role**

You are a 35 year old woman.

You have been having pains and numbness in your right hand for about a month. The pain is in the thumb and first two fingers of your right hand mostly, but also goes a little way up your arm. The numbness and tingling is just in your thumb and fingers.

You have no problems in your other joints.

When the problem first started, it was just in your fingers and usually came on when you had been using your hand to lift something or had been using your hand a lot. It has gradually got worse and worse and is now almost constant. It gets worse if you are using your hand a lot – typing on the computer, cooking, washing – all those household things. The pain sometimes wakes you up at night.

You do find that shaking your hand in the air relieves the pain a little but it is no longer enough. You have been taking 8 paracetamol a day for the pain – this just about keeps the pain manageable so you can continue to use your hand.

You just want to find out what it is and get it sorted.

Background

You are a University lecturer in mathematics.

You have diabetes, for which you take metformin every day. You are careful with your diet. Your blood sugar is well controlled – you see the diabetic nurse regularly who checks your HbA1C. The levels have been only just above the normal range for a while now.

You do not have any allergies to any medicines that you are aware of. You have no other health problems and have not gained weight recently. You are not overweight.

Your parents are alive and well. You have two brothers and one sister; your sister, an aunt and an uncle all have diabetes.

You are married with two daughters. Your husband and children are well.

If asked if you could be pregnant, you would hope not as you have had as many children as you want and have had your tubes tied.

You have never smoked and you do not drink alcohol because it is against your religion.

N.B. Allow the doctor to examine your hands. Everything should be normal except that if the doctor

- taps on your wrist under the base of the thumb, or
- asks you to put your hands together back-to-back, fingers pointing downwards

this brings on the numbness and tingling.

45 History/Explanation   **Examiner mark sheet**

Introduces self and role and checks identity of patient

Explains/confirms purpose of interview (hand pain)

- site
- onset
- character
- radiation
- associated symptoms elsewhere (shoulder, neck problems)
- exacerbating/relieving factors (especially shaking it in the night)
- severity
- previous occurrences of the problem

Establishes effect on daily life (makes household tasks difficult)

Explores ideas, concerns, expectations (just wants it to get better)

Drug history (including over the counter and illicit drugs)

Allergies

Establishes previous medical history,

- Especially diabetes, thyroid disease, rheumatoid arthritis, recent weight gain, pregnancy (all possible precipitating factors)

Establishes family history

Establishes Social history

- occupation
- who is at home
- smoking
- alcohol

Excludes other systemic symptoms

Brief examination to include Phalen's and/or Tinel's test

Explains what the problem is likely to be – Carpal Tunnel syndrome

Uses an explanation the patient would understand (nerves trapped beneath the fibrous band at the wrist)

Gives advice for managing the problem:

- resting the hand – no using the right hand for lifting, washing up, typing, for example
- use of a splint or wrist support (can be bought from the chemist or refer to physiotherapist for one), including at night
- possibly injection or surgery if problem continues – so return if it is not improving in six weeks.

Appropriate questioning technique (mixture of open and closed questions)

Avoids or explains jargon

Uses tools such as signposting and summarising

Systematic, logical approach

Checks if the patient has any other questions

Finishes consultation appropriately

Friendly approach and appropriate body language

45 Notes

46 History   Candidate role

You are a foundation doctor in General Practice.

Your next patient is a 59 year old woman who rarely sees the doctor.

Please take a history.

At the end tell the examiner the differential diagnoses.

46 History   Patient role

You are a 59 year old woman.

You don't know what is wrong at the moment, you just do not seem to be yourself. You are so tired, you feel as weak as water. And your neck and shoulders ache. They are really stiff in the mornings, it takes you a while to get going. You can't understand why they should be like this – it is not as if you have been doing anything unusual. You do like to do the garden but you have done that for years.

You have been feeling this way for about five or six weeks on and off.

You wonder if you might be depressed – maybe retiring last year was not such a good idea? You had hoped to get out and about with the local walking club as you love walking, but even though you would like to go out you just have not got the energy. Thank goodness for the telly, that seems to be all you are up to it these days. You do enjoy the nature programmes and Gardeners' World.

Background

You have lost a couple of pounds – even your appetite seems to be less these days. You are sleeping OK and your mood is – well, fed up but not so bad.

You do not smoke, or drink alcohol.

You rarely take exercise (except taking your grandson out for a walk in his pushchair).

You are a retired schoolteacher and live with your husband.

46 History   Examiner mark sheet

Introduces self and role and checks identity of patient
Explains/ confirms purpose of interview
Establishes nature of presenting complaint (as relevant)
- tiredness
- onset
- pattern
- exacerbating and relieving factors

Enquires about relevant associated symptoms:
- Features suggestive of depression
    o Mood
    o Loss of interest/ pleasure in life
    o Sleep
    o appetite
- Features suggestive of systemic disease
    o Weight loss
    o Pain – site, onset, character, radiation, time pattern, exacerbating and relieving factors
- Excludes other systemic symptoms
    o Bowel habit
    o Waterworks
    o PV bleeding

Establishes effect on daily life
Establishes patient's concerns, validates and addresses these (depression – the patient is fed up rather than depressed, she still enjoys some activities even if she is frustrated)
Drug history (including over the counter and illicit drugs)
Allergies
Establishes previous medical history
Establishes family medical history
Establishes Social history
- occupation
- who is at home
- smoking
- alcohol

Appropriate questioning technique (mixture of open and closed questions)
Avoids or explains jargon
Uses tools such as signposting and summarising
Systematic, logical approach
Checks if the patient has any other questions
Friendly approach, appropriate body language
Finishes consultation appropriately

46 Notes

Differential diagnoses

Polymyalgia rheumatica
Depression
Cancer (unknown site)
Vitamin D deficiency
Another cause of arthritis

> LEARNING POINT
> This history is typical of a patient with polymyalgia rheumatica, but notice how easy it would be to misdiagnose it as a case of depression, especially as the patient herself is thinking of this possibility. The fact that she still enjoys certain activities goes against depression as a diagnosis. Missing the morning shoulder girdle stiffness in this case could mean you fail to check an ESR and start this patient on high dose steroids. Without rapid treatment this patient could go blind if her polymyalgia is associated with giant cell temporal arteritis.

> LEARNING POINT
> Be wary of patients with vague non-specific symptoms such as tiredness and be prepared to look for a physical or psychosocial cause
> Balancing the risk of missing a diagnosis with the risk of over-investigating an unhappy or stressed patient is one of the skills of general practice and is a good reason for continuity of care in the community. A doctor who knows the patient well will know if this is a patient who comes to the doctor a lot, with every ache and pain, or if it is unusual for her to complain and really is likely to signify an underlying change in her health.

Obstetrics and gynaecology practice scenarios

47 History   Candidate role

You are a Foundation doctor in General Practice.

Your next patient is a 19-year-old student who has come to the surgery asking for the pill. Please take a history and advise her as needed.

47 History   Patient role

You are a 19 year old student.

You have come to the GP for the pill. You have not been on the pill before. You have been using condoms with your boyfriend over the last three months, but do not want to rely on them. You have had no condom accidents so far but you don't really feel safe. And you don't plan to use them once you are on the pill as they kind of interrupt sex...(although you might not tell the doctor this).

Your last period was ten days ago. Your cycle is regular. You usually bleed for 4-5 days once a month.

Background

You are normally fit and well. You do get occasional headaches but nothing serious. Your headaches are behind your eyes and across your temples. You have to sit quietly in a dark room several hours. Ibuprofen and co-codamol help.

You have never been in hospital and have not had any serious illness that you remember.

You live with your parents and go to college. You smoke ten cigarettes per day.

If the doctor agrees to prescribe the pill you may want to know when you can start it. You are fed up with condoms...

47 History and explanation   Examiner mark sheet

Introduces self and identifies role

Checks patient's identity

Confirms/ establishes reason for visit

Asks about current contraception

Asks about LMP and possibility of current pregnancy

Enquires about past contraception, especially if patient has been on pill before

Enquires about menstrual cycle
- regularity
- any period problems (pain/ heaviness/ inter-menstrual bleeding)

Past Medical History - asks specifically about contraindications to the pill
- DVT/PE;
- focal migraines (a full description of the headaches must be obtained to be sure that they are not migraines)
- breast problems (benign breast disease is not a contra-indication to the pill, cancer is)
- serious liver problems
- high blood pressure
- BMI >35 relative contraindication

Drug History – enzyme inducing drugs interact with the pill

Social history – smoking/ alcohol (older smokers should not be on the pill – give health education now)

Explanation

Establishes patients level of knowledge of pill

Explains that no contraception is 100% effective but if 100 women take the pill correctly for a year less than 1 will become pregnant on the pill

Explains how to take the pill
- every day for 3 weeks then one week break
- start with first day of next period (can start straight away if she will use condoms for the first week and is certain she has not had any missed condoms this month)
- within 12 hours of same time every day, it is best to make it a regular part of the day – set an alarm, or do it every morning when you brush your teeth for example
- still covered for contraception during break
- explains that missing pills increases the risk of getting pregnant accidentally, especially if the missed pills are close to the week's break – advises patient to read the leaflet in the pack about missed pills
- suggests condoms in addition for STD protection

Explains potential side effects

- Rarely can cause clots in the leg, so if she has pain or swelling of her calf she should see a doctor straight away and tell them she is on the pill
- Any other problems make an appointment to discuss
- If there is no bleed in the week's break she should check she is not pregnant

Establishes patient's concerns, validates and addresses these

Uses chunking and checking - giving small pieces of information and checking understanding before continuing

Checks if patient has other questions

Uses communication tools, such as signposting and summarising, appropriately

Offers written advice or websites

Language appropriate throughout

Fluent and professional manner

Friendly approach with appropriate body language

47 Notes

LEARNING POINT

There is no evidence that taking the oral contraceptive pill in the early stages of pregnancy causes harm to the foetus, but many doctors prefer to advise patients to start the pills with the first day of the next period for simplicity's sake – that way you know for certain she is not pregnant..

Do always check if the patient needs emergency contraception when she consults for the first time about contraception.

Missed pill advice can be quite complicated. A simple way of explaining the risk regarding missed pills to a patient is to advise that the pill puts the ovaries to sleep. Seven days is not long enough for the ovaries to wake up and release an egg. Eight or nine days may be enough time, so around the week's break is the most risky time to miss pills.

---

48 Explanation   Candidate role

You are a Foundation doctor in General Practice.

Your next patient is a 40-year-old lady who has come to the surgery to discuss contraception.

48 Explanation   Patient role

You are a 40 year old woman.

You are attending the surgery because you want to start contraception and would like the GP to advise you. You have not really thought about what you would like to use.

You have 3 children and have never used contraception. Your husband was always careful. He died two years ago and you are in a new relationship. You have met a 35-year-old divorcee but you are unsure that it will last. You have not had sex yet but you want to be prepared and you definitely do not want to get pregnant.

Background

You had your first period when you were 14 years, and your last period 7 days ago. Your cycles are regular and last 5 days per month.

You had been married for 6 years before having your first child. Your children are 9, 6 and 4 years old.

You are normally fit and well.

If the doctor asks, remember that you did have a clot in your leg after your last pregnancy, but that is all sorted now. You took warfarin for it – you remember now because it was a hassle getting your blood checked all the time.

You think your mother had a problem with a clot too. The doctor said it was a clot that caused her stroke. She died five years ago.

You smoke about 10 cigarettes a day and drink 5 units of alcohol per week.

48 Explanation   **Examiner mark sheet**

Introduces self and identifies role
Checks patient's identity
Confirms/ establishes reason for visit
Enquires about current contraception
Asks about LMP and possibility of current pregnancy
Enquires if patient has chosen preferred method
Enquires about past contraceptive history
Enquires about menstrual cycle

- regularity
- any period problems (pain/ heaviness)

Past medical history, including specifically asking about contraindications to the pill

DVT/PE, breast problems, liver problems, migraines, diabetes, high blood pressure

Drug history (including non-prescribed drugs)
Explains options, using written information as an aid and discussing pros and cons of each

- The combined pill is contraindicated by her past history of DVT. Her age and smoking status are also relative contraindications for the combined pill.

A combination of methods might be a good idea in this situation, i.e. condoms plus:

- POP
- Depo provera injection
- Nexplanon
- IUCD
- Or possibly condoms alone with advise about seeking the morning after pill if needed?

Finds appropriate option that woman is happy with
Uses chunking and checking - giving small pieces of information and checking understanding before continuing
Establishes patient's concerns, validates and addresses these
Checks if patient has other questions
Uses communication tools, such as signposting and summarising, appropriately
Offers written advice or websites
Language appropriate throughout
Fluent and professional manner
Friendly approach with appropriate body language

48 Notes

LEARNING POINTS
A first contraception consultation is always a long one if the patient is to be given adequate information. That is why having written information as a back up is so useful. The patient may not wish to make a choice there and then but can take a leaflet away to think about it. Giving condoms meanwhile is a good idea.

Excellent summary leaflets are available on www.fpa.org
Look for the leaflets that outline all the choices.

49 History   Candidate role

You are a Foundation doctor in General Practice.

Your next patient is a 20-year-old woman who has come to the surgery to discuss contraception.

49 History   Patient role

You are a 20-year-old married woman.

You are attending the surgery because you are frightened of getting pregnant again.

You had an abortion when you were 16, and you have 2 children, aged 3 years and 6 months. Both your children are in foster care - social services took them. You are not likely to want to talk about what happened unless you really feel the doctor is OK. In fact your partner, Patrick, was accused of hitting the kids and social services also asked you if he had abused them sexually. You know that he has a temper but you don't believe he is a pervert. You hope to get the babies back - that much you will tell the doctor.

You were given the pill after the second baby, but Patrick wants you to have another, so he flushed them down the toilet. He also beat you up but you would only admit to this is you really trust the doctor.

You do not want another child right now as you know that you need to sort out your life first. You need something that is guaranteed to work and that Patrick will not know about. You want to start it today and you are not keen on talking about all the other stuff.

Background

Your first period was when you were aged 13 years, and your last period was 2 weeks ago.

You had (unprotected) sex two nights ago.

Your cycle is regular and lasting 3-4 days in a month.

You are normally fit and well and have never been admitted in a hospital.

You live with Patrick, who drinks a few beers most nights. You drink sometimes but not as much as him, maybe just a vodka and orange if you go out (and that is rare these days). You don't do drugs.

49 History   **Examiner mark sheet**

Introduces self and identifies role

Checks patient's identity

Confirms/ establishes reason for visit

Enquires if patient has chosen preferred method (something partner will not find out about which she can start today)

Enquires about current contraception

Asks about LMP and possibility of current pregnancy

Enquires about menstrual cycle

- regularity
- any period problems (pain/ heaviness)

Enquires about past contraceptive history

Past medical history

Drug history

Offers suitable options, discussing pros and cons of each

- emergency contraception (levonorgestrol/ulipristal/IUCD)
- Depo Provera – patient would ideally be given post-coital contraception, then could have injection today but the change to menstrual cycle would be a disadvantage as partner may be aware
- Nexplanon (contraceptive implant) - patient would ideally be given post-coital contraception, then could have injection today but the change to menstrual cycle would be a disadvantage as partner may be aware, he may also feel it in her upper arm
- IUCD – could be fitted up to day 19 of cycle to act as emergency contraception and ongoing contraception, with threads cut short partner should not know about it, can make periods heavier or more painful

Finds appropriate option that woman is happy with

Elicits history of domestic violence

Offers appropriate support in view of domestic violence

- discusses safety – does she want to stay with him? Does she have somewhere she can go to if she wants to escape? Can she keep a bag with a friend just in case?
- helpline number / social worker / return visit

Uses chunking and checking - giving small pieces of information and checking understanding before continuing

Establishes patient's concerns, validates and addresses these

Checks if patient has other questions

Uses communication tools, such as signposting and summarising, appropriately

Offers written advice or websites

Language appropriate throughout

Fluent and professional manner

Friendly approach with appropriate body language

49 Notes

LEARNING POINTS

It is estimated that as many as one in four women suffer violence from their partner at some time in their lives. It is easy to under-estimate the control that a violent partner may exert. Once the woman's self-esteem is destroyed, she can be manipulated into staying on in a relationship that may lead to physical injury, mental illness, substance abuse and even death. You should not be afraid to ask a woman if she feels safe in her relationship, and to offer help should she report problems. Women in abusive relationships may not accept help the first time it is offered but by ensuring that the woman knows there is a safe exit route, you can make it easier for her to seek help later.

See

www.rcgp.org.uk/policy/position_statements/domestic_violence-the_gps_role.aspx

50 Explanation

Candidate role

You are a Foundation doctor in gynaecology outpatients.

The next patient, a 29 year old woman, has been referred for tubal ligation. Please discuss this with her.

50 Explanation   **Patient role**

You are a 29 year old woman.

You are married with three children aged five, three and six months.

You and your husband have decided that you do not want any more children. You do not think that you could cope – the baby cries all night, the three year old always gets into your double bed in the small hours and the five year old still wets the bed. You gave up work to bring up the children and now can't wait until they are old enough for you to go back to the office.

You have seen your GP about getting sterilised and have now come up to clinic to see the hospital specialist.

You have no special worries. You know the operation is done through that new camera method.

Your best friend had a sterilisation five years but her marriage broke up and now she is with a new partner. She is due to have a reversal of the sterilisation procedure next month (with BUPA – a private medical company).

Background

At the moment your husband is using condoms, but neither of you really like this – you are worried that it is not really safe enough.

You never liked the pill because it made you put on weight and gave you headaches. You are afraid of needles so you don't want to have the injection. And you don't like the idea of a coil. So you feel sterilisation is the best way.

Your GP had suggested a vasectomy with your husband, but when you discussed it with him, he was quite squeamish and worried about the effect it might have on your sex lives. Since you don't mind the tube-tying operation, it seems the easiest.

50 Explanation   **Examiner mark sheet**

Introduces self and identifies role

Checks patient's identity

Confirms/ establishes reason for visit

Establishes how many children/ pregnancies patient has already had

Establishes current methods of contraception and contraceptive history

Enquires what patient's partner thinks of her being sterilised

Establishes what patient knows about sterilisation, correcting any misconceptions

Discusses irreversibility clearly

Discusses failure rate (1/200)

Describes what will happen (before during and after procedure)

- general anaesthetic
- small cut below umbilicus
- tubes clipped or cut or burned
- small cut at bikini line
- no stitches
- day case surgery – home after tea and toast
- slightly bruised and tired for a few days

Describes risks (minor, common and rare but serious)

- anaesthetic
- bruising
- bleeding
- infection
- conversion to laparotomy

Advises patient to continue contraception until following period

Advises periods will not be affected (women on the pill may notice their periods change – this is because they are coming off the pill, not as a result of the operation)

Uses chunking and checking - giving small pieces of information and checking understanding before continuing

Establishes patient's concerns, validates and addresses these

Checks if patient has other questions

Uses communication tools, such as signposting and summarising, appropriately

Offers written advice or websites

Language appropriate throughout

Fluent and professional manner

Friendly approach with appropriate body language

50 Notes

LEARNING POINTS

In the UK, only the patient needs to consent for sterilisation, however it is good practice to enquire about both parties' views. If a couple disagree on sterilisation, there is more likely to be regret afterwards, which may lead to requests for reversal. Reversal is rarely available on the NHS and success is not guaranteed as the tubes remain scarred and the ciliary transport mechanisms that waft an egg along the tube will never return to normal.

Many gynaecologists also counsel against sterilisation immediately after a delivery or termination, and in women under thirty, again on the grounds that decisions made at times of change or crisis are more likely to lead to regret later on.

51 Explanation   Candidate role

You are a Foundation doctor in General Practice.

Your next patient is a 23 year-old lady who is 12 weeks pregnant. She was with the midwife for her first antenatal check, but the midwife was called away before she could answer all the patient's questions.

Please see her and discuss her concerns.

51 Explanation

Patient role

You are a 23 year-old married woman and have recently found out you are pregnant. You have been referred to the midwife and are due to see her in a couple of weeks, but you want to see the doctor today as you are quite excited and nervous about having this baby.

You would like to know what tests are available, so that you can know in advance if there is a problem with the baby. Your husband is very supportive but is also worried about the possibility of having a handicapped child. If there is something very serious, you and your husband have decided that you would have an abortion

And you want to know if there is anything else you need to do to make sure the baby is healthy?

Background

You have been reading the literature given to you by your midwife and are worried about having a handicapped baby. Your sister had a child with cerebral palsy last year and she finds it very hard to cope sometimes, especially as her husband is not very helpful. Her marriage is under a lot of strain. You don't want that to happen to you. You want to have all the tests, but you are not sure what is on offer.

You smoke and drink alcohol. You have cut down on both since finding out you are pregnant. You are down to three cigarettes a day and only take one or two glasses of wine per week.

You are not on any medicine other than paracetamol for the occasional back pain, and have never been in hospital before.

You work in a post office and plan to go back to work after having the baby as you need the money.

51 Explanation

Examiner mark sheet

Introduces self and identifies role

Checks patient's identity

Confirms/ establishes reason for discussion

Asks if patient wants anyone with her

Finds out the patients' concerns and addresses these appropriately
- elicits history of CP in family
- explains that cerebral palsy is not something that is known to run in families
- explains that there are no specific tests for CP

Explains what tests are available
- regular check-ups, including blood pressure (to make sure it is not too high), urine test (to check for protein and sugar) and abdominal examination (to make sure the baby is growing appropriately)
- blood test to check for anaemia and to check her blood type
- blood test to check rubella immunity
- blood test for infections - hepatitis, syphilis, HIV
- blood test and USS for risk of Downs (further tests offered if risk is high)
- USS for abnormalities such as spina bifida, serious heart problems

Explains that even with normal tests there is no guarantee of a normal baby and the tests carry some risks

Explains the things patient can do to reduce risk to baby
- take folic acid until 16 weeks pregnant as it reduces the chance of spina bifida
- always tell doctor or pharmacist you are pregnant before taking new medicines
- give up smoking
- keep alcohol down to two glasses of wine per week max – nil is best
- avoid children with obvious signs of infectious illness
- do not change pet litter (as it can carry infection)
- avoid smelly cheeses, runny egg yolks and strong pates (foods linked with potential for food poisoning)

Offers follow up with midwife in case of further questions

Uses chunking and checking - giving small pieces of information and checking understanding before continuing

Establishes patient's concerns, validates and addresses these

Checks if patient has other questions

Uses communication tools, such as signposting and summarising, appropriately

Offers written advice or websites

Language appropriate throughout

Fluent and professional manner

Friendly approach with appropriate body language

51 Notes

52 Explanation

Candidate role

You are a Foundation doctor in general practice.
Your next patient is a woman aged 24.
Please deal with her problems today.

52 Explanation

Patient role

You are a 24 year old woman.

You have come to see the GP today for a referral for sterilisation.

You are single and do not want children.

You are a PA to a company director and are determined to climb the career ladder. You want to move to London to work with a bigger firm next year to increase your prospects.

Last month you had a termination of pregnancy. You fell pregnant after knowing the man for only three weeks and that was the last you saw of him! You never want to go through that experience again.

You can't take the pill because it makes you put on weight and gives you headaches.

You are afraid of needles so you are not going to have the injection.

And you have had pelvic inflammatory disease – you had to be admitted to hospital for antibiotics and painkillers – and the doctor then told you should never have a coil.

Men cannot be trusted to use condoms – that is how you got pregnant.

You know that sterilisation is irreversible and that is what you are wanting.

You will be quite irritated if the doctor starts to argue with you. You know your mind and do not accept that you are too young to make this kind of decision (which is what your friend told you when you discussed it with her).

52 Explanation — Examiner mark sheet

Introduces self in friendly manner

Confirms identity of patient

Establishes patient's reason for attending (for referral for sterilisation)

Establishes how many children/ pregnancies patient has already

Establishes current method of contraception and contraceptive history

Explores ideas, concerns and expectations around sterilisation

Discusses irreversibility clearly

Discusses the issue of the patient being very young to make such a big decision that she may regret later:

- Some surgeons will not operate of someone of this age.
- Alternatives are available (Mirena coil is highly effective and does not carry a risk of PID; implant is also highly effective and lasts for three years)

Able to keep discussion going in a calm manner

Offers either

- more time to think (with information on alternatives) and come back to talk again
- or referral for further discussion with gynaecologist.

Asks if patient has any questions

Understands patient or asks for clarification

English easily understandable throughout

Uses language appropriate to the patient throughout

Friendly approach, appropriate body language

Maintains an appropriate level of eye contact throughout

Fluent, professional manner throughout

52 Notes

LEARNING POINT

This station raises the difficult issue of patient and doctor autonomy. The Mental Capacity Act states that we must respect patient autonomy; that patients have the right to choose a path that we may think will not lead to the best clinical outcome. But doctors also have the right to step away from treatments that they feel may be harmful, so whether or not a gynaecologist agrees to sterilise this patient might depend on how likely the doctor felt it would be that the patient would regret the decision later.

In this consultation you have to keep the doctor patient relationship alive, give clear advice about alternative options and document carefully the discussion.

How would you feel about this station if the patient were carrying a genetic trait that she does not want to pass on? Would this alter your discussion? Remember that the same options still exist.

53 History   Candidate role

You are a Foundation doctor in General Practice.

Your next patient is a 25-year-old woman, who is concerned about her fertility.

Please take a relevant history from her, and advise her.

53 History   Patient role

You are a 25 year old woman.

You have come to the GP because you have been trying for a baby and you have not managed to get pregnant yet.

You have been married for three years. Before that, you had not had a sexual relationship. For the first couple of years of marriage you were on the pill, because you were finishing a course in accountancy. You stopped the pill and have been trying for a baby for about 8 months now. You have never been pregnant.

You and your husband have sex three or four times a week. This is full sex – your husband puts his penis into your vagina.

Your last period was ten days ago. Your cycle is regular. You usually bleed for 4-5 days, once a month.

Background

You are normally fit and well. You have very mild asthma, for which you have a blue inhaler that you need to use once or twice a week. You have never been in hospital and have not had any serious illness that you remember. You do not take any regular medication.

Your parents are both well. You have two sisters and one brother. You do not know of any problems with fertility in your family.

You live with your husband and work for a local college as a junior accountant. You do not smoke and have stopped drinking alcohol since you have been trying for a baby.

Your husband is 30 years old and is a rugby coach.

He is very healthy. He has had knee injuries in the past, but he has no ongoing illnesses and does not take any regular medication.

He often goes out drinking at the weekend, and sometimes in the week. You are not sure how much he drinks regularly, but you have seen him drink several pints of beer without getting drunk, so you suspect it is quite a lot.

He also smokes, but he doesn't smoke in the house because you don't like it, so you are not sure how many cigarettes a day.

You know he has had a long-term partner in the past, but he has no children.

53 History   **Examiner mark sheet**

Introduces self and identifies role
Checks patient's identity
Confirms/ establishes reason for visit
Discusses relevant main points of history:
- how long they have been trying for a baby
- use of any contraception in the past
- any previous pregnancies?
- menstrual cycle
- how often the couple have penetrative sex
- use of any other medication
- past medical history, especially gynaecological (including operations and pelvic infections)
- fertility problems in the family
- smoking and alcohol use

Discusses partner's relevant history:
- any pregnancies of previous partners?
- known health problems (including operations or treated STDs)
- smoking and alcohol use

Explains relevant tests and/or treatment:
- not being pregnant after eight months of trying is not unusual
- would consider investigations if no pregnancy after 1 year (earlier if there were physical problems)
- there are tests and interventions if couple have problems

Gives general recommendations to improve chances of conception:
- be sure to have sex two or three times per week during the middle of the cycle (explain clearly when this is)
- both partners to have a healthy diet
- both partners reduce or stop smoking and alcohol intake

Gives general preconception advice
- folic acid 400mcg daily
- could consider checking rubella status

Uses chunking and checking - giving small pieces of information and checking understanding before continuing
Establishes patient's concerns, validates and addresses these
Checks if patient has other questions
Uses communication tools, such as signposting and summarising, appropriately
Offers written advice or websites
Language appropriate throughout
Fluent and professional manner
Friendly approach with appropriate body language

53 Notes

54 Explanation   Candidate role

You are a Foundation doctor in General Practice.

The next patient is a 29 year old woman, who is attending today asking to be referred for an abortion.

54 Explanation   **Patient role**

You are a 29 year-old lawyer who is in her first pregnancy.

You are about six weeks pregnant. Normally your periods are regular but the last one was six weeks ago. You have done about ten home pregnancy tests and they all say the same.

You are looking for promotion at work which your boss will not give you if you have a baby. You are married. You husband doesn't know you are pregnant and neither does your family. You were using condoms for contraception.

At the moment you feel your promotion is more important than having a baby. You would like to be referred for an abortion.

You don't know much about what is involved as you had quite a strict Catholic upbringing, but you are sure that it is what you want to do.

Do you need to tell your husband?

Do you need to tell work?

You are otherwise very fit and well.

54 Explanation   Examiner mark sheet

Introduces self and identifies role

Checks patient's identity

Confirms/ establishes reason for visit

Establishes gestation of pregnancy

Establishes reasons for wanting termination

Discusses briefly alternatives for the patient (keeping the baby or adoption)

Establishes understanding of procedure

Explains two types of abortion possible, medical or surgical

- Medical termination is carried out at about eight or nine weeks pregnant. It involves:
  - o taking a tablet to block the pregnancy hormones
  - o then a couple of days later having a pessary inserted into the vagina
  - o the patient usually goes home and has a very heavy bleed
  - o most patients will need to take painkillers
  - o a few patients will need to go into hospital for strong pain relief or to have a small operation
- Surgical termination is carried out a few weeks later at about 10 – 12 weeks. It involves:
  - o having a general anaesthetic
  - o the womb is scraped out
  - o the procedure is done as a day case

Asks if discussed with anyone/ who is supporting her

Encourages patient to someone close to her for support

Reassures patient that the consultation is confidential

Stresses that they will support her whatever her choice

Offers referral or to pass the patient to another doctor who will refer.

Uses chunking and checking - giving small pieces of information and checking understanding before continuing

Establishes patient's concerns, validates and addresses these

Checks if patient has other questions

Uses communication tools, such as signposting and summarising, appropriately

Offers written advice or websites

Language appropriate throughout

Fluent and professional manner

Friendly approach with appropriate body language

54 Notes

LEARNING POINTS

Make sure you know the law around termination of pregnancy in the UK. Although in theory we do not have termination on demand, in practice doctors will almost always sign to allow a woman, who is in the first trimester of pregnancy, to have a termination on the grounds of protecting her physical or mental health. At later stages of the pregnancy, availability of termination becomes more restricted (although still legal).

If your personal beliefs mean that you do not wish to be involved in termination of pregnancy, you must still be aware of the law and of the guidance of the GMC. The GMC advises that doctors must 'not withhold information about the existence of a procedure or treatment because carrying it out or giving advice about it conflicts with your religious or moral beliefs...you must ensure that arrangements are made, without delay, for another doctor to take over their care. You must not obstruct patients from accessing services or leave them with nowhere to turn. Whatever your personal beliefs may be about the procedure in question, you must be respectful of the patient's dignity and views.'

General Medical Council Personal beliefs and medical practice - guidance for doctors, March 2008.

55 History   Candidate role

You are a Foundation Doctor in General Practice.
Your next patient is a 36 year old woman.
Please take a history from her.

At the end of the consultation please state
- your differential diagnoses
- the investigations you would like to perform.

55 History   Patient role

You are 36 years old and female.

You have pain in your stomach and your motions have been a bit loose. You are worried in case you have given yourself food poisoning from a reheated takeaway.

Background

The pain is on the right side of your stomach, just below the level of your tummy button. It is a bit like a period pain, except it has been there for three days – normally you only get the period pain on one the first day of your period. The only other thing you have noticed is that you have started to feel a bit sweaty and a bit lightheaded over the last day or so.

You have not noticed any change in your waterworks. You think your motions have been a little bit loose in the last few days. You have not had any other health problems with this pain. You have not eaten anything unusual recently, except you heated up and ate the leftovers from the previous night's takeaway four nights ago (a curry). You have been a bit off your food since then, though you have not been sick.

Your periods are usually regular, you bleed for two or three days every 30 days. You often get some mild period pain for about a day. Your last bleed was three weeks ago, but it was a lot less than usual – hardly anything. You do not use contraception as you would like to have children. You have had a slight brownish discharge over the last couple of weeks.

You have plantar fasciitis of your right foot, which was diagnosed two years ago (painful sole, no cause found – you take painkillers and do exercises that were shown to you by a physiotherapist). You do not take any other medication.

You work as a sales assistant in a bookshop. You are married, but do not have any children. Your parents are both well, though your mother takes tablets for an underactive thyroid. You like to drink bacardi breezers and tend to have quite a few at the weekend, but not during the week – maybe five or six on Friday and Saturday nights. You do not smoke.

55 History   Examiner mark sheet

Introduces self and role and checks identity of patient
Explains/confirms purpose of interview
Establishes nature of presenting complaint (as relevant)

- site
- onset
- character
- radiation
- time duration
- exacerbating and relieving factors
- severity

Enquires about relevant associated symptoms:

- weight loss?
- change in appetite?
- diarrhoea?
- blood (how much and when)?
- Urinary symptoms (pain, blood, frequency)
- Gynaecological (LMP, normal cycle, discharge, contraception/chance of pregnancy)

Establishes effect on daily life
Explores ideas, concerns, expectations (concern over food poisoning)
Drug history (including over the counter and illicit drugs)
Allergies
Establishes previous medical history
Establishes family medical history
Establishes Social history

- occupation
- who is at home
- smoking
- alcohol

Excludes other systemic symptoms
Appropriate questioning technique (mixture of open and closed questions)
Avoids or explains jargon
Uses tools such as signposting and summarising
Systematic, logical approach
Checks if the patient has any other questions
Finishes consultation appropriately
Friendly approach and appropriate body language

55 Notes

Differential diagnoses

Ectopic pregnancy
Ovarian cyst
Appendicitis
Pelvic inflammatory disease
Gastroenteritis

Investigations

Urine pregnancy test
PV with swabs if pregnancy test negative
FBC (anaemia, white cell count for infection)
Ultrasound scan
(Laparoscopy)

> LEARNING POINT:
> Never forget the possibility of pregnancy or a related complication in a woman of reproductive age.

56 History   Candidate role

You are a Foundation doctor on-call for gynaecology. It is 9pm.

Switchboard bleep you, and when you answer they put you through to a woman who is concerned because she is pregnant and bleeding.

Please find out more about the problem.

Deal with any concerns that she has and suggest a course of action for her to take.

56 History   Patient role

You are 29-years old, married to James.

You are 10 weeks pregnant and have noticed that you have some spotting of blood on your pants this evening. You have not had any tummy pain with the bleeding.

This is your second pregnancy. You are due to see the midwife for the first time next week.

Your first pregnancy, 2 years ago, ended in a miscarriage at about the same time, so you are quite worried and have telephoned the hospital to ask for advice. You want to know if there is anything that you can do to stop the same thing (miscarriage) happening again. Are there any tablets you could be given?

You have no medical problems that you know of and the only medicine you are taking is folic acid.

Your periods are quite irregular and the time between periods can vary from 24 to 33 days. When your period comes, you bleed for two or three days.

56 History

Examiner mark sheet

Introduces self and role and checks identity of patient

Explains/confirms purpose of interview

Establishes nature of presenting complaint (pv bleeding)

- Checks this is pv not pr bleed
- onset
- character (fresh red or brown old blood? Clots?)
- severity – how much blood
- duration of pregnancy
- previous scans this pregnancy

Enquires about relevant associated symptoms:

- abdominal pain
- discharge
- fever
- symptoms of UTI, other illnesses and any problems with the pregnancy

Past obstetric history – outcome of past pregnancies

Drug history (including over the counter and illicit drugs)

Allergies

Establishes previous medical history

Explores ideas, concerns, expectations

Explanation

What is happening

- Informs the patient that miscarriage is possible, but not inevitable – many women bleed in early pregnancy but go on to have a normal baby

What should be done

- Advises rest tonight
- See GP in the morning, as if the bleeding continues a scan can be then arranged
- No need to be seen in hospital tonight unless the bleeding:
  - o gets worse
  - o is accompanied by bad pain
  - o there is a fever or other new symptoms

Checks that the patient has understood the information, is happy with advice and has the ability to get further help if needed (safety netting)

Appropriate questioning technique (mixture of open and closed questions)

Avoids or explains jargon

Uses tools such as signposting and summarising

Systematic, logical approach

Checks if the patient has any other questions

Finishes consultation appropriately

Friendly approach

56 Notes

57 History   Candidate role

You are a Foundation doctor in obstetrics.

Your next patient has been admitted with a small vaginal bleed at 33 weeks gestation.

Please take a history from her.

At the end of the consultation, tell the examiner
- your differential diagnoses
- your management plan

57 History

Patient role

You are a 31 year old lady in her second pregnancy.

Your periods were irregular before you got pregnant. According to the midwife and the scan you are about 33 weeks pregnant, and the baby is due in 7 weeks. You only had one scan and that was at about 10 weeks, when they said the baby was fine. You were supposed to have another one later but you were on holiday and missed that appointment.

You had a bit of a fright this morning when you lost a bit of blood – it was like the start of a period – fresh red blood on your pants- you had to use a pad. You went to the GP straight away and he sent you to the hospital.

You are very frightened that something may go wrong with the baby. You think that you have felt the baby kicking but to be honest you have been so worried that you are not sure what is happening. You have not had any pain and have not had any problems before this is the pregnancy.

Background

Your first pregnancy ended in a miscarriage when you were about three months pregnant, which was why you had the early scan this time round.

You work as a secretary and your husband is in a sales assistant.

You are a non-smoker and have not taken any alcohol since you found out you were pregnant.

You took folic acid tablets in the beginning of your pregnancy but have not been prescribed anything since. You do take a Pregnacare vitamin supplement you buy from the chemist.

You have not been in hospital since you had your tonsils removed as a child. You are allergic to elastoplast (a type of sticking plaster).

57 History   Examiner mark sheet

Introduces self and role and checks identity of patient
Explains/confirms purpose of interview
Establishes nature of presenting complaint:
- patient description of problem
- severity (how much blood?)
- onset (when did it start?)
- duration (how long? is she still bleeding?)
- pain
- foetal movements
- has it happened before?

Obtains details of the pregnancy
- Last menstrual period (LMP)
- Expected due date (EDD)
- problems earlier in the pregnancy?
- scans?

Brief past obstetric history (ask about other pregnancies or any operations)
- Gravidity and Parity
- Outcome of pregnancies

Explores ideas, concerns, expectations (fear for baby)
Drug history (including over the counter and illicit drugs)
Allergies
Establishes previous medical history
Establishes family medical history
Establishes Social history
- occupation
- who is at home
- smoking
- alcohol

Excludes other systemic symptoms
Appropriate questioning technique (mixture of open and closed questions)
Avoids or explains jargon
Uses tools such as signposting and summarising
Systematic, logical approach
Checks if the patient has any other questions
Finishes consultation appropriately
Friendly approach and appropriate body language

57 Notes

Differential diagnoses

Placenta praevia
Placental abruption
Vasa Praevia
Incidental bleeding from cervix or vaginal lesion

Management plan

Examine the abdomen
IV access
FBC
Group and save/ cross match
USS
NB: Do NOT do a vaginal digital examination at this stage – this might cause a major bleed from a placenta praevia

58 History

Candidate role

You are a Foundation doctor in General Practice.
Your next patient is a 22-year old woman.
Please take a history of her presenting complaint
At the end of the history tell the patient what you wish to do next and what you think is the most likely diagnosis

58 History   Patient role

You are a 22-year old woman.

You have come to the doctor today because of an embarrassing problem. Your fanny has become very itchy and red and you get a bit of thick, white discharge. Having sex has become painful (the pain is on the outside) and so has going to the loo. This all started about five days ago. You are embarrassed to talk about this and will only mention the smell of the discharge and the pain during sex if asked directly.

You are not peeing more than usual and everything else is normal.

Background

Your periods are regular and the time between periods is just less than a month (it starts earlier each month by two or three days). When your period comes, you bleed lightly for three or four days. Your last menstrual period started last week (on Tuesday) and lasted for the usual time.

You have no other health problems and are not taking any regular medication except the Pill for contraception, though you did have some antibiotics from your GP for a cough. You finished them just over a week ago. They were called something like amsicill. The Pill you take is called Microgynon and you take it every day.

You work as a waitress in a café.

You have a regular boyfriend who you sex with. You've been together for the last three months. You have had several boyfriends in the past. Your boyfriend has told you that he has had sex with other girlfriends before he met you.

You smoke about 10 cigarettes a day and an occasional joint and you have about 5 or 6 drinks on a Friday or Saturday. Bacardi and coke is your favourite, but you'll drink most things, except beer. You don't use recreational drugs (that's hard stuff like heroin, isn't it?).

Your parents are in good health (though your father takes tablets for blood pressure). You have a younger sister who has diabetes and injects herself with insulin, she has had the diabetes since she was 9.

You have no idea what has caused this problem. You just want to get it sorted out.

58 History   Examiner mark sheet

Introduces self and role and checks identity of patient
Explains/confirms purpose of interview
Establishes nature of presenting complaint:
* patient description of problem
* duration
* odour of the discharge
* itch
* has the patient noticed any associated features (abdominal pain/ pain on intercourse)
* aggravating/relieving factors/anything the patient has tried to help
* previous occurrence of problem

Asks about likely relevant features:
* urinary problems
* menstrual problems
* LMP
* sexually active?
* does the partner have symptoms?
* contraceptive use

Establishes effect on daily life
Explores ideas, concerns, expectations
Drug history (including over the counter and illicit drugs)
Allergies
Establishes previous medical history (antibiotics for recent infection)
Establishes family medical history
Establishes Social history
* occupation
* who is at home
* smoking
* alcohol

Excludes other systemic symptoms
Appropriate questioning technique (mixture of open and closed questions)
Avoids or explains jargon
Uses tools such as signposting and summarising
Systematic, logical approach
Checks if the patient has any other questions
Finishes consultation appropriately
Friendly approach and appropriate body language

58 Notes

Plan

Vaginal examination with triple swabs (high vaginal swab, endocervical swab and chlamydia swab)

Likely diagnosis

Thrush secondary to antibiotics

Treatment

Clotrimazole pessary +/- cream, or oral fluconazole 150mg stat

---

LEARNING POINTS

Until 2010, doctors were advised to tell women on the oral contraceptive pill that broad spectrum antibiotics might interfere with its action. The standard advice was to use condoms during the course of antibiotics and for one week afterwards. This is no longer considered recommended as the evidence suggests it is not necessary unless the antibiotics are strong enzyme inducers such as rifampicin.

Common risk factors for thrush include:
 antibiotics
 pregnancy
 raised blood sugar
however the majority of cases probably have no recognised precipitant.

Clotrimazole cream may dissolve the latex of condoms. Patients who are reliant on condoms for contraception must be warned to abstain or use alternatives whilst using clotrimazole cream.

It is not usual to treat the male partner of a woman with vaginal thrush unless the problem is persistent, as it is not considered to be sexually transmitted in the majority of cases.

Paediatric practice scenarios

59 History   Candidate role

You are a Foundation doctor answering phone calls for the out of hours service in the middle of the night. A parent phones up for advice about their six year old child, Jessica, who has a fever and a painful neck.
Please take the phone call.
Take a full history and advise the parent on further management.

When you have finished the consultation please tell the examiner your differential diagnoses and why you chose to manage the patient as you did.

59 History   **Patient role**

It is 10pm. You are the single parent of Jessica aged six.

You have just come in from your evening shift in the supermarket to find your daughter is ill. Your mother has been looking after her. She is concerned because Jess has a temperature and is complaining of a poorly neck. Jess hardly ate anything at tea-time because she says her neck hurts too much when she swallows. She is whingey tonight, complaining of her neck. You know that a painful neck can be a sign of meningitis, so you have phoned the doctor.

You want the doctor to visit (and will be quite insistent on this). You do not want to go into the primary care centre because you think it might be dangerous for your daughter to go out in the cold when she is ill.

Background

Jess is usually well. She hardly has a day off school. She is on no medicines and has never been in hospital. You have just given her a teaspoonful of ibuprofen before phoning the doctor.

Jess is your only child.

You think that she may have a slight rash – she is a bit pink with some small pimples. **If the doctor asks you to press a glass on the spots, tell them the spots go white**.

She watched TV on the sofa all evening and played her computer games. She can put her chin onto her chest and kiss her knees (if asked to by the doctor).

She has not complained of leg pains, and her hands and feet do not feel cold. She has not turned blue and has no other symptoms.

You have not taken Jess's temperature as you do not have a thermometer, but she feels hot to you when you out a hand on her forehead.

59 History   Examiner mark sheet

Introduces self and role and checks identity of patient/ relative
Confirms purpose of interview
Establishes nature of presenting complaint
- Character of symptoms – clearly distinguishes between sore throat and stiff neck – asks if child has pain on swallowing; is able to put her chin on chest and/or kisses knees
- Duration

Associated symptoms
- temperature/ feeling hot all over/ hands and feet cold?
- rash – obtains clear description, colour, site, result of tumbler test
- asks about activity level/ consciousness level
- eating?
- vomiting?
- diarrhoea?
- tummy pain?
- earache?
- cough?
- asks about infectious contacts
- enquires what parent has been doing to manage case

Past medical history and drug/ allergy history
Establishes Social history - who is at home
Explores parents ideas, concerns, expectations (worry over meningitis)
Appropriate questioning technique (mixture of open and closed questions)
Management plan
- reassures parent appropriately that it does not sound like a meningococcal rash
- advises allowing ibuprofen time to work then reassess, continuing antipyretics if needed (but not exceeding recommended doses)
- advises most sore throats do not need antibiotics
- ensures that parent is happy with advice
- offers seeing child in out of hours centre if parent still worried – explains that going out in cold is not a danger to child
- safety nets – advises of danger symptoms/ to be seen if worse

Avoids or explains jargon
Uses tools such as signposting and summarising
Systematic, logical approach
Checks if the patient has any other questions
Finishes consultation appropriately
Friendly approach, appropriate body language

59 Notes

Differential diagnoses

- Viral infection/URTI/pharyngitis/tonsillitis
- Meningococcal disease unlikely as child active / able to kiss knees/ rash blanching, but need to safety net by encouraging parent to call again if symptoms change and child becomes sicker.

60 History/advice   Candidate role

You are a Foundation doctor in the Accident and Emergency department.

A mother has brought in an 18 month old child, Jasmine, who is reluctant to put weight on her left leg. You have examined Jasmine and found that she has fresh purple bruising on her thigh and also bruising to her right ear, old yellow bruises on her left arm and what appear to be tooth marks in her right upper arm.

Please discuss with the mother what has happened and what will happen now.

At the end of the consultation the examiner will ask you what your differential diagnoses is and why, and who will be involved in the further management.

60 History/advice

Patient role

You are the mother of 18 month old Jasmine.

You have brought her to A&E because she is limping. You are not sure how this happened – your boyfriend, Darren, was looking after her yesterday whilst you went to see your sister in hospital. He told you that she had fallen down the stairs. You are not sure whether or not to believe him as you know that he has got a bad temper at times – but you would not tell this to the doctor.

Background

Jasmine has never been to hospital before. She has had all her vaccinations and has only had the normal coughs and colds. She started walking a few steps when she was one and she can say a few words, like Mam, juice etc.

If the doctor suggests that Jasmine is admitted you are likely to make excuses as to why you cannot stay. You are worried because Darren did not want you to go to bother the doctor. He said that Jasmine was just playing up for attention and would not let you call anyone yesterday when she was clingy. He does not know that you have brought her to hospital today.

Darren is not Jasmine's Dad - her Dad did not stick around when you found out you were pregnant. Darren can get quite angry with her when she makes a lot of noise but you know that deep down he loves you both and would not do anything to hurt her , don't you?

60 History/advice   Examiner mark sheet

Introduces self and role and checks identity of patient

Explains/ confirms purpose of interview

Establishes nature of presenting complaint, probing gently:
- to elicit mother's idea of what happened
- who witnessed events
- when mother noticed a problem and what she did
- reason for delaying presentation

Elicits social history
- who is at home
- relationship of Darren to Jasmine

Past medical history/ medication/ vaccinations/ development
- Asks about previous medical problems and visits to the doctor or hospital
- Asks about vaccination history
- Asks about developmental milestones

Establishes mother's concerns, addresses and validates these (worry about how it happened and about what Darren will say if he finds out she has gone to the doctor)

Explanation – what will happen now
- Explains that the child needs admission to hospital for further investigations
- Gently probes the reasons for the mother's reluctance for the child to be admitted
- Is able to keep the mother calm but does not allow the child to go home
- Reaches an agreement with the mother over the next steps
- Involves seniors (may mention referral to child protection services/ social services but may leave this discussion to more senior colleagues)

Checks if patient has questions

Uses communication tools, such as signposting, summarising, chunking and checking appropriately

Language appropriate throughout (avoids jargon)

Fluent and professional manner

Friendly approach, appropriate body language

60 Notes

Differential diagnosis

Non-accidental injury – pointers include
- multiple injuries of different ages
- the bite mark
- the delay in presenting an injured child
- the history of a non-related partner who has a temper and was the only adult present at time of injury.

Differentials might include accidental injury and clotting or skin disorders that predispose to bruising, but these could not account for the bite.

Those who will be involved

Senior paediatricians
Child protection team/ social services
Possibly the police and the courts

61 Explanation   Candidate role

You are a Foundation doctor in the Accident and Emergency department.
The ST4 has asked you to speak to the parent of Euan, a 20 month old who
was brought in after a 3 minute convulsion. All investigations have been
normal and the diagnosis is febrile fit during a probable viral upper
respiratory infection.

Please give Euan's parent the necessary advice before discharge.

You do not have to take a history for this station

61 Explanation

Patient role

You are the parent of Euan, who is 20 months old.

Yesterday Euan had a bit of a snotty nose and was off his food but otherwise he was OK. Then this morning he had a fit. You were just getting him up when it happened. It was terrifying. It probably only lasted for a couple of minutes but it felt like an age. You wrapped him in a blanket and brought him straight to Casualty.

Euan has been very thoroughly examined by the doctors (and had blood and urine specimens taken) and now one of them has come to speak to you. You are feeling a bit better, because Euan is now looking happier, but you still have some concerns:

You want to know what caused the fit

Does it mean he is epileptic like your sister-in-law?

You are glad that Euan is in hospital where he can be observed by professionals (you will not feel ready to take him home just yet).

Supposing he has another fit? What would you do? You would be very worried about dealing with it and as you are on your own at home there is no-one else to help.

61 Explanation   Examiner mark sheet

Introduces self and identifies role

Checks patient's identity

Confirms/ establishes reason for visit

Establishes parent's understanding of what has happened to Euan

Establishes patient's concerns, validates and addresses these (fears of epilepsy; what to do if he fits again)

Facts to include in the explanation:

What has happened

- Serious diseases, such as meningitis, have been ruled out and the diagnosis is that Euan had a febrile fit

What this means

- Febrile fits are common in children under the age of five
- They do not mean that Euan has any long-term problem;
- Children who have febrile fits only rarely go on to develop epilepsy (2/100)
- There is a 1/3 chance that Euan will have a further fit if he has a temperature in future, but simple fits are not dangerous

What can be done about it

- If he does have another fit, lie him on his side and keep him safe from injury. Do not force anything into his mouth. Call for help if the fit lasts longer than five minutes.
- If he is ill, treat him as usual. Give paracetamol or ibuprofen if he is distressed but be aware this will not prevent febrile fits.
- Parents of children who repeated have febrile fits may be given rectal diazepam to use if a prolonged fit occurs.
- Euan should still have vaccinations as normal

Uses chunking and checking - giving small pieces of information and checking understanding before continuing

Checks if patient has other questions

Uses communication tools, such as signposting and summarising, appropriately

Offers written advice or websites

Language appropriate throughout (avoids jargon)

Fluent and professional manner

Friendly approach, appropriate body language

61 Notes

62 Explanation   Candidate role

You are a Foundation doctor in General Practice.

Your next patient is Whitney, a three year old little girl.

You see in the records that she has just been treated by the out of hours doctor for a urine infection. This is the third time she has had a urine infection.

Her parents were advised to make an appointment at her General Practice and are here to see you now. Please discuss with them what will happen next.

62 Explanation   **Patient role**

You are the parent of Whitney, your three year old daughter. You have come to see the GP because the out of hours doctor told you to make an appointment.

At the weekend Whitney was very feverish and complaining of pain when she was weeing. You called NHS direct and they told you to call the GP. When you did that the doctor told you to bring her in to the treatment centre. There he examined Whitney and tested a sample of her wee. He said it showed she had a water infection. He gave her some medicine to take twice a day and told you to go and see the GP during the week. He said that Whitney may need some more tests. So here you are.

Whitney is much better now. Her temperature has settled and she is not going to the toilet so often.

Whitney has had this problem a couple of times before but then the doctor just gave her antibiotics and she got better. You are not clear why you have been told she now needs more tests.

Background

Other than the water infections she has not had any serious illnesses in the past, just the usual coughs and sore throats.

She has had all her vaccines and goes to nursery. You have a son of seven who is fit and well.

You (or your wife if you are male) have had urine infections a couple of times in the past as an adult, but the doctor just treated you with antibiotics. You did not have to have any tests. So why is this necessary for your daughter?

62 Explanation   **Examiner mark sheet**

Introduces self and identifies role

Checks identity of parent and child

Confirms/ establishes reason for visit

Establishes understanding of situation

Establishes how Whitney is now

Facts to include in the explanation of the need for investigation:

- One off infections in children are not uncommon and do not need further tests as long as the child gets better.
- Repeated urine infections in small children are usually investigated as they can sometimes be a sign that there is a problem with the urinary tract.
- Explains that referral to the paediatrician will be made so that the specialist can decide which investigations are needed

Discusses briefly possible investigations:

- Further urine culture to make sure infection has cleared
- blood test (U&Es)
- USS
- special X-rays involving a dye injection so that the kidneys can be seen working (DMSA or micturating cystogram).

Offers to see the child again if unwell meanwhile

Establishes patient's concerns, validates and addresses these (urine infections in children can be more serious than in adults so may need more investigation)

Uses chunking and checking - giving small pieces of information and checking understanding before continuing

Checks if patient has other questions

Uses communication tools, such as signposting and summarising, appropriately

Offers written advice or websites

Language appropriate throughout (avoids jargon)

Fluent and professional manner

Friendly approach, appropriate body language – reassuring and calming manner

62 Notes

LEARNING POINT
To pass this station you do not need to know the exact details of which investigations will be carried out, but you do need to know:

-that a small child with repeated urine infections should be investigated
-the kind of investigations that may be used
-how to advise the parents of this in a reassuring manner.

See NICE Guidance CG54 for information about how to investigate infants and children with UTIs:. www.guidance.nice.org.uk/CG54/.

63 Explanation

Candidate role

You are the on-call Foundation doctor in paediatrics.

A six month old baby, Michael, was admitted earlier with bronchiolitis. He is propped up on head box oxygen in his cot, but is on no other medication. The parent of Michael has called you over for a word. Please answer her questions as best you can.

63 Explanation

Your son, Michael, aged six months, has been admitted to hospital.

He has had a cough for a few days. You took him to the GP, but were told that it was just a virus and to give him calpol. He seemed to get worse and worse and you were worried because he could barely take his bottles, so you brought him to the children's hospital.

The casualty doctor has told you that he needs to stay in hospital overnight to have oxygen. He says it is an infection in his airways.

You are very worried about Michael and very angry that your GP did not give him any antibiotics or anything. You want to talk to the doctor about this.

Background

Michael is your first child. You are an insurance broker. Your mother usually looks after Michael whilst you and your partner are at work.

You smoke ten cigarettes per say – but never in front of Michael.

63 Explanation   **Examiner mark sheet**

Introduces self and identifies role

Checks patient's identity

Confirms/ establishes reason for visit

Establishes patient's concerns, validates and addresses these (worry for Michael and belief that GP missed the diagnosis)

- Acknowledges the stress of the situation
- Avoids criticising colleagues

Facts to include in the explanation:

- cause of bronchiolitis (Respiratory Syncytial Virus)
- antibiotics do not work as RSV is a virus
- most children with bronchiolitis can be managed at home with simple measures as suggested by the GP, hospital admission is only needed for the worst cases
- children with severe bronchiolitis may be given oxygen and a special feeding tube to help them whilst they recover.
- Michael is likely to have to stay in hospital for a few days until he is better

Uses communication tools, such as signposting and summarising, chunking and checking appropriately

Checks if patient has other questions

Offers written advice or websites

Language appropriate throughout (avoids jargon)

Fluent and professional manner

Friendly approach, appropriate body language

63 Notes

> LEARNING POINT
>
> This station tests your ability to deal with angry relatives and well as your understanding of bronchiolitis. Use this as a reminder to revise the General Medical Council's guidance on working with colleagues
> http://www.gmc-uk.org/guidance/good_medical_practice/working_with_colleagues.asp
> Whilst you should not be afraid to question colleagues if you feel that patient care would be improved, you should be careful not to undermine the trust patients have in them by careless remarks or unfounded criticisms.

Psychiatry practice scenarios

64 History   Candidate role

You are a Foundation doctor in General Practice.

Your next patient is a 52 year old woman, complaining of low mood.

She was diagnosed with fibromyalgia (joint and muscle pains with no abnormality found on investigation) three years ago, which has been treated with a variety of things - exercise, amitriptyline, tramadol and over-the-counter painkillers at different times.

Please take a history from her about her mood.

At the end of the consultation please tell the examiner
- The diagnosis
- Whether this patient poses a suicide risk
- The treatment options you would consider most likely to help her.

Note

Your patient appears clean and well-dressed.

64 History   Patient role

Note: do not smile during the discussion – you don't feel like smiling.
You are a 52 year old woman.
You have been diagnosed as suffering from fibromyalgia for the last four or five years. This gives you aches and pains almost all over and makes you very tired. You have had all sorts of tests but nothing shows up. And you have tried all sorts of treatments; exercise, amitriptyline, tramadol and over-the-counter painkillers at different times. Nothing really gets rid of the pain fully.
At the moment you are taking tramadol three or four times per day. Usually you also swim twice a week to help the fibromyalgia, but you haven't been able to motivate yourself to go the pool for a few weeks.
You have come to see the doctor because the pains in your arms have been worse again and your husband says you have 'not been yourself' for some time. He thinks you might be depressed. You think he might be right, but are more concerned that you are going to have these pains forever.
You have also started to get headaches several times a week, lasting for a couple of hours at a time. You take paracetamol for the headaches, but don't feel it helps very much.
If asked by the doctor:
- You feel unhappy most of the time, but the mornings are the worst. You would like to just stay in bed and pull the blankets over your head, which is ironic because you spend most to the night tossing and turning, watching the clock.
- You get off to sleep but then wake in the small hours and can't get back to sleep.
- Your appetite is OK, you eat well, but only through habit, not enjoyment. You don't think you have lost or gained weight recently
- You have lost interest in life and can't enjoy anything. You are an art teacher and used to enjoy your job, but now it is all you can manage to get the class working on a project. You seem to have just lost your normal drive. Even your own art work has lost its appeal, you just can't be bothered now.
- You would rather avoid friends than go out socialising as you used to.
- You have completely lost interest in sex which isn't helping your relationship with your husband.
- You feel very tired most of the time, and you keep bursting into tears over nothing at home.
- You are just feeling useless really, like nothing is ever going to get better, which is silly really because you know you have nothing to complain about compared to some people.
- You are finding work very difficult (physically and mentally) and have had several days off sick in the last few months

If the doctor asks you if you have thought about harming yourself or committing suicide, you might admit to having thought your husband would be better off without you, but if pushed you would say you have never really made any plans. Your religion would stop you. You are Catholic.

You would also be shocked if your doctor asked if you have heard voices or seen things that weren't there – that only happens if you're going mad, doesn't it? You don't think you've had any memory problems – remind the doctor that you are only 52 if they ask you about your memory – you are too young for that to be going.

Background

Your husband is a joiner, but is finding it hard to get much work at the moment, so your wage is important right now. You have three children, they are all OK. The last of your children to leave home left in September last year to go to Bolton University.

You rarely drink and then it is only a gin and tonic on high days and holidays. You are a non-smoker.

You have been getting thoroughly fed up with your fibromyalgia. The only things that almost always help are hugging a hot water bottle and relaxation therapy, but you don't always have the opportunity to do the techniques – usually you want to be able to get on with your life, not lie around thinking about beaches.

Maybe you are depressed? You don't know but enough people have suggested it to you. What does this doctor think? Is there anything that will help you to feel better?

64 History   Examiner mark sheet

Introduces self and role and checks identity of patient
Explains/ confirms purpose of interview
Establishes nature of presenting complaint
- mood
- diurnal rhythm
- sleep (distinguish between early morning wakening and difficulty getting off to sleep)
- appetite
- enjoyment of usual pleasures
- concentration/ motivation
- feelings of shame or worthlessness
- delusions/ hallucinations
- suicidal thoughts/ plans/ protective factors

Establishes effect on daily life
Explores ideas, concerns, expectations (depression; continuing joint pains)
Drug history (including over the counter and illicit drugs)
Establishes previous medical history including previous mental health problems/ pre-morbid personality
Establishes Social history
- occupation
- who is at home
- smoking and alcohol

Appropriate questioning technique (mixture of open and closed questions)
Avoids or explains jargon
Uses tools such as signposting and summarising
Systematic, logical approach
Checks if the patient has any other questions
Finishes consultation in suitable manner
Friendly approach with appropriate body language

64 Notes

Diagnosis is moderate depression.
This patient does not seem to be high risk of suicide as she has not made any plans and has protective factors in her religion and her thoughts of her family.
The treatments to consider here would include
- Cognitive behavioural therapy – this has been proven to be beneficial in depression and can also help some patients suffering long term painful conditions such as fibromyalgia.
- Drug treatment – probably using a selective serotonin re-uptake inhibitor as first line as they have a good profile in terms of patient acceptability.

65 History   Candidate role

You are a Foundation doctor on call for psychiatric admissions to accident and emergency. You have been called to see a patient, a 32 year old woman, who has taken an overdose. Medically she does not need to be kept in hospital.

Please take a history.

At the end of the history discuss the key factors that might influence your decision as to whether or not this woman can go home.

65 History   **Patient role**

You are a 32 year old woman.

You took half a bottle of vodka and 12 paracetamol tonight as you have simply had enough. You cannot face going on. You would be better off dead. You took the tablets whilst you were alone at home. You had told your Mum that you were having a girls night out and had asked her to babysit Jemma, your three year old. So you did not think anyone would find you. Your Mum found you when she called in to the house to pick up extra nappies for Jemma. She brought you to hospital when you told her what you had done.

You had written a letter to your Mum, asking her to look after Jemma for you for good. She would be a better Mum than you anyway and has been doing most of the childcare because you have been working.

You have been working in a call centre to try and get a bit of cash together but you hated your job. People are always rude to you and you find it very stressful. You had several days off work because you felt simply too stressed to come in and the bosses were getting quite threatening about all your time off. Last week you burst into tears and ran from the centre when someone was rude. Now you cannot face going back in and so you called them a couple of days ago to hand in your notice. Yesterday the job centre told you that you could not get any money because you had resigned.

You have been sleeping OK, except for when Jemma wakes you up, or when the local yobs on the estate kick off in the street. Except last night, you had a bad night after the job centre business.

If it weren't for Jemma you would not have got this far. You still love Jemma and she is the one thing that gives you pleasure any more.

You do your best to keep your flat nice but it is difficult on your own, plus the window was broken by some lads down the street a few weeks ago. The council promised to get it mended but no-one has been so far. The cold comes in and the wooden board you put up lets the damp in. You must be a rubbish mother if you cannot even provide your child with a good home.

You do not have a partner. He disappeared during the pregnancy.

You have not been out with your friends for a long time. It is almost impossible with child care arrangements and you really do not feel like it anyway. If you do see anyone, you do not want to sit and moan – they do not want to hear all your problems. You would just make them miserable, and you have no spare cash, so why go out?

The health visitor used to call round when Jemma was small, but you have not seen her for ages. She did come one time asking about postnatal depression but she asked loads of questions of a form and you really do not like that kind of thing, so you said you were alright.

You have never taken an overdose or tried to harm yourself before.

You used to be quite a bright and bubbly kind of person at school. It is just since then that everything has gone wrong. You don't know what happened to that person.

You can't see things getting better for you in the future. Life is just miserable. But you might be surprised if the doctor asks you about hearing voices or any odd things like that.

You have never been in hospital before. You had mild asthma as a child but have not been on any medication for years. No-one in your family has ever had to see a psychiatrist as far as you know.

You do not normally drink much alcohol except when you go out with friends and that has not been for ages now. You do smoke about ten cigarettes per day but try not to do this around the baby.

You just want to be left alone. You feel quite stupid for what you have done – you even failed with that! You are not sure whether or not you would do it again.

One of the doctors in casualty took some blood from you a while ago and told you that the results show you do not need any treatment, but he would not discharge you home until you had spoken with the psychiatry doctor. The psychiatry doctor has come to see you now.

65 History   **Examiner mark sheet**

Introduces self and role and checks identity of patient
Explains/confirms purpose of interview
Establishes details of the overdose

- timing
- number of tablets taken
- alcohol
- place and who was present (likelihood of being found)
- letter
- present view of overdose

Establishes events leading up to overdose

- who is at home
- social support, family and friends
- work situation
- financial situation

History of depression

- mood
- diurnal rhythm
- sleep (distinguish between early morning wakening and difficulty getting off to sleep)
- appetite
- enjoyment of usual pleasures
- concentration/ motivation
- feelings of shame or worthlessness
- delusions/ hallucinations
- pre-morbid personality
- past history of self-harm/ involvement with mental health services/ mental health problems

Establishes previous and family history of mental illness

- smoking
- alcohol and drugs

Past medical history
Appropriate questioning technique (mixture of open and closed questions)
Avoids or explains jargon
Uses tools such as signposting and summarising
Systematic, logical approach
Checks if the patient has any other questions
Finishes consultation appropriately
Friendly approach and appropriate body language

65 Notes

This patient planned the suicide attempt but chose a low pain method.

She still says that she would be better off dead, although if she can be helped with her practical problems (housing, finances, child care) this may change.

She has social support and a reason to live (for her daughter). She still enjoys caring for her daughter. She has no past history of attempted suicide.

All of these factors together suggest she might be safe to discharge *provided* support can be put in to get her over this crisis and you are happy that the care for her daughter is adequate.

An appropriate discharge plan might include

- discharge to care of mother
- crisis team to offer support and follow up over the next couple of weeks
- GP to be involved (prescribing antidepressants)
- Health Visitor to help find child care

LEARNING POINTS

Factors which suggest a high suicide risk include:

Suicide method planned (not impulsive), and tried to avoid being found

Violent methods (hanging, jumping) higher risk than pain free methods (overdose)

Past history of serious (almost successful) attempts

Physical or social problems which CANNOT be overcome (e.g. terminal diagnoses) and therefore induce a feeling of hopelessness

Severe mental illness

Continued statement of intent

Lack of social support / social crises

Older age

Male> female

---

66 History   Candidate role

You are a foundation doctor in psychiatry.

Please take a history of alcohol use from your next patient, who has been referred by their GP because of a drink problem.

66 History

Patient role

You are 34 years old.

You used to work for a building company, but are now unemployed.

Your GP has sent you to see the alcohol team, because he is concerned about your drinking. You also feel that it is a problem – you know you drink too much. You want to give up if only the doctors can give you something to stop the shakes and stop you wanting the booze.

You have drunk alcohol since you were a teenager – a few cans of beer a week at first. At 16 you first started going into pubs with your mates, but you still only drank two or three pints of beer a week or a couple of snakebites! Then you got a job (with the building company) and you could afford more. You used to just drink on a Friday and Saturday evenings. Gradually, the drinking extended through the week as well as the weekend.

You found you often had rows with girlfriends about it – they always nagged that you were more interested in spending time with your mates than with them, and though you had relationships, they never lasted long.

By your late twenties, most of your mates had got married and weren't available to go out drinking, so you either went out by yourself to your local pub, or stayed in and watched the telly with some cans of lager. You felt very depressed at this time as everyone else seemed to be moving on, having families and you were not. It was easier to drink to forget everything.

You started having problems at work because you could not get into work on time. Eventually, you had a big row with your boss and walked out before he could sack you. This was five years ago. You have been on the social ever since, so money is tight.

Now it just something that you feel you have to do: drink. As cider is the cheapest, you started drinking this at home. It is cheaper than the pub. Now the cider is the only alcohol you drink. Currently you drink about three bottles of White Lightning cider a day (6 litres), every day.

You feel awful if you don't have a drink – you have to have a drink in the morning otherwise you get sweaty and shaky and feel like you want to puke. You don't go without alcohol for more than a few hours if you can help it. You did have a fit once and ended up in hospital for the night, when you tried to stop drinking for a couple of days.

You sleep badly, never have much energy and feel depressed, though you like to watch football on the telly. You don't have much hope for the future – you've really cocked it up haven't you? Who would want an alcoholic as a friend or partner? You came along to see the alcohol team today because your GP insisted, but you don't think they will really be able to do anything for you. You don't see your family often, because your mum and dad nag you about your drinking and you just end up having a row about it.

If asked, you have never been in trouble with the police because of alcohol and although you tried cannabis when you were younger, you do not take any street drugs now. You smoke roll ups if you can cadge the tobacco.

66 History   **Examiner mark sheet**

Introduces self and role and checks identity of patient
Explains/ confirms purpose of interview
Establishes nature of presenting complaint
- What is drunk
- How much, pattern across day and across the week
- How long this pattern has been established
- Past attempts at stopping drinking – results

Use of other substances of abuse (illicit drugs/ tobacco)

Establishes effect on daily life/ Problems caused by drinking:
- Relationships
- Work
- Finances
- Police
- Housing
- Health/ accidents/ black outs/ fits/liver problems

Mood (depression screening)
Explores ideas, concerns, expectations
- Reasons for wanting help now
- What help he wants from medical profession
- How confident he feels in his ability to give up now
- What will stop him or make it hard to give up

Establishes previous medical history, including drugs and allergies
Establishes family medical history
Establishes Social history
- occupation
- who is at home
- drinking habits of social circle

Excludes other systemic symptoms (eating, vomiting, indigestion, bowels, jaundice)
Appropriate questioning technique (mixture of open and closed questions)
Avoids or explains jargon
Uses tools such as signposting and summarising
Systematic, logical approach
Checks if the patient has any other questions
Finishes consultation in suitable manner
Friendly approach and appropriate body language

66 Notes

LEARNING POINTS
Students may include recognised screening questions in their history. These are set out below for information purposes. However it is important to distinguish a screening test from a full history in a patient known to have alcohol problems. This patient is already known to have harmful drinking habits, so a simple screening test is insufficient.

Patients do not talk in alcohol units. You must take a history of what is drunk and when, probing to ensure that all the alcohol taken is included (e.g. beer/cider/wine and spirits). Then you will need to convert this into approximate units - there are tools to help you do this. A full history is more important than the ability to do the conversion in your head.

Have you ever...[5]

C	Felt you needed to cut down
A	Been annoyed by others criticising your drinking
G	Felt guilty about your drinking
E	Needed an eye opener

Audit - Shortened screening questions[6]
How often do you have a drink containing alcohol?

| Never | Monthly | 2-4 / month | 2-3 / week | >3 / week |

How many units of alcohol do you drink in a typical drinking session?

| 1-2 | 3-4 | 5-6 | 7-9 | >9 |

How often do you have more than 6 (f) or 8 (m) in one drinking session?

| Never | <Monthly | Monthly | Weekly | More than weekly |

Questions are scored (0-4) and a complete AUDIT questionnaire completed if the respondent scores more than 3/12

FAST test[7]
1. How often do you have eight or more drinks on one occasion?
2. How often during the last year have you been unable to remember what happened the night before because you had been drinking?
3. How often during the last year have you failed to do what was normally expected of you because of your drinking?
4. Has a relative or friend, a doctor or other health worker been concerned about your drinking or suggested you cut down?

Answering yes to any one of these questions should prompt a more complete history.

---

[5] Mayfield D, McLeod G, Hall P (1974). The CAGE Questionnaire: Validation of a New Alcoholism Screening Instrument. The American Journal of Psychiatry; 131(10): 1121-1123.

[6] The Alcohol Use Disorders Identification Test. Guidelines for use in primary care. Babor TF, Higgins-Biddle JC, Saunders JB & Monteiro MG. (2001). World Health Organisation. Available at www.whqlibdoc.who.int/hq/2001/WHO_MSD_MSB_01.6a.pdf

[7] Manual for the Fast Alcohol Screening Test (FAST). Health Development Agency and University of Wales College of Medicine.
Available at http://www.nice.org.uk/niceMedia/documents/manual_fastalcohol.pdf

67 History   Candidate role

You are a Foundation doctor in Emergency Medicine.

A confused elderly lady (Emily James, aged 83) has been brought in by a friend.

Please take a history from the friend.

After your history be prepared to describe
- your differential diagnoses
- the next steps you would take, with justification for why you think they are important.

67 History   Patient role

You are a neighbour of Mrs Emily James aged – 80ish.

You called an ambulance for your neighbour this evening because, when you went round to see her, she did not seem like her normal self. She seemed confused, all sort of groggy. You were very worried, so you called the ambulance.

Background

You usually call in to see her once or twice a week. You normally ask if she wants anything from the shops when you go, although she is quite active and very independent. She likes you to get her heavy vegetables as she still cooks a proper dinner for herself every day.

You last saw her three days ago, when nothing seemed wrong.

Normally she does the crosswords in the paper – she has a very active mind. You know she has a daughter who lives in Chesterfield and you got her phone number but you could not get her on the phone this evening.

You know that she is on a number of tablets but you do not know what they are. She had mentioned getting her heart check at the doctors but you don't know any details. Now you are regretting not looking for her tablets whilst you waited for the ambulance, but you did not like to pry.

She had wet herself when you found her and she was quite upset about it – you will only admit this if asked directly about how she looked. You helped her to change her pants before the ambulance came. She was a bit smelly which is unusual for her.

You are a bit worried about getting home yourself. You have had to get your sister to babysit the kids and you know she can't stay late. You really need to get away soon.

67 History   Examiner mark sheet

Introduces self and role and checks identity of the friend
Explains/ confirms purpose of interview
Established events immediately prior to admission
- how was patient found?
- any clues as to what had happened?
- when was she last seen beforehand?
- other witnesses?

Established how patient is normally (i.e. level of functioning normally)
Past medical history
Drug history (prescribed and recreational)
Social history
- living circumstances
- occupation
- alcohol
- smoking

Asks directly about anything else relevant not already mentioned:
- falls injury?
- bottles of pills?
- recent illnesses/ infections

Allows for questions
Appropriate questioning technique (mixture of open and closed questions)
Avoids or explains jargon
Uses tools such as signposting and summarising
Systematic, logical approach
Checks if the patient has any other questions
Finishes consultation appropriately
Friendly approach and appropriate body language

67 Notes

Differential should include:
- infection (UTI/ chest)
- drugs (prescribed)
- electrolyte imbalance (e.g. from anti-hypertensives)
- head injury from fall
- intracranial lesion (bleed or neoplasm)
- other physical illness (silent MI / thyroid)
- psychotic illness

The next steps would be to
- examine the patient looking for signs of infection/ injuries/ smells/ signs of drug use...
- investigations including Glucose, FBC, U&Es, Calcium, CRP, MSU, CXR, CT/ MRI head (after discussion with the radiologist)

LEARNING POINT

Issues of confidentiality and consent here – because the patient lacks capacity, the doctor should act in the patient's best interest. In taking a history from a friend or witness we are not breaching confidentiality. It is quite acceptable to give general information such as, 'we will be keeping her for a while and doing some tests'.

---

68 History   Candidate role

You are a Foundation doctor in General Practice. Today, a slim 17 year old female studying for her 'A' levels has come to see you, brought by her mother who is outside in the waiting room.

Please take a history
- Discuss the possible diagnosis with the patient.
- Advise her of the next steps in your management plan.

68 History

Patient role

You are 17 years old and female. You are still at school studying for your A-levels. You want to go on to do languages at university and are planning to take a year out working in France before then.

Your Mum has forced you to come to the doctor. You really do not want to be there (and you may not give him all the information that he wants, if you do not feel you want to talk openly). Your mum was going to come in to the consultation with you but you made her stay in the waiting room. She has been pestering you about your weight for months. You know that she thinks you are too thin, but you disagree. She is just jealous because she cannot lose weight. She has always said that she would like to lose a few pounds, and you do not want to end up overweight like her.

You have been dieting for a while. It started when you were sixteen and wanted to lose enough to look good in your bikini in the summer. All your friends admired your will power and you found that dieting was quite easy for you. You have a lot of will power.

You now weigh around 44kg (6 stone ten pounds), and are not actively trying to lose weight now but you do not want to be any heavier. At your heaviest you weighed 58kg. You are 5 foot 6inches tall.

You do not want to look fat and you know that your bum is still gross. You would argue that you are just eating healthily – you eat fruit and vegetables and are vegetarian. You skip breakfast if you can (though you probably will not admit this to the doctor). You take black coffee and will pretend to your mum that you have had some cereal or toast, she is not around when you get up most days. You usually have a diet drink for lunch and a piece of fruit. You like to drink pepsi max to keep you going and can get through four cans a day. The evening meal is always a problem if you can't get out of it. There is often a row about you picking at your food. You try to stick to the vegetables. You might have an apple later in your room when you are studying to hold off the hunger pangs.

You do sit ups in your room at night – you can manage about 100 at a time. You do not make yourself sick very often, but you do take laxatives. You have been taking ex-lax – up to a packet at a time, two or three times per week. The disadvantage is that they give you terrible stomach cramps, but at least they flatten your tummy.

You are a good student and are doing well at school. Your mock exam results give you predicted A grades. You are looking forward to getting away from home to get your mum off your back and be independent. France should be great.

You have two sisters, one older, one younger. Your older sister is overweight. She is away at university and has put on even more weight since she has been cooking for herself. When she came home last summer she was dieting and your mum did not complain about her. Your younger sister is a skinny little girl and eats all sorts of rubbish.

Your periods have never been regular. You only get one every three or four months, but that does not really bother you.

You have never been in hospital and are generally fit and well.

Your mood is OK. You are a bit tense about your exams but then so is everyone at this stage. You are sleeping OK – although you do get up early to revise and do your sit ups.

You do not have a boyfriend, but have a few friends at school. You do not smoke and only drink a glass of wine at parties. You have tried a joint once but have never used other drugs.

Remember, you do not feel there is a problem and you are likely to get quite irritated by too much questioning, unless the doctor is really gaining your trust. You certainly would not agree to a diagnosis of anorexia if the doctor tried to discuss it with you.

68 History   Examiner mark sheet

Introduces self and role and checks identity of patient

Explains/confirms purpose of interview

Elicits the patients' understanding of the situation (mum thinks she is too thin)

Elicits the patient's ideas, concerns and expectations (wants to be left alone)

Asks about:

- eating habits - goes through daily eating routine including meals, drinks and snacks in between
- current weight and height, and any changes in weight
- how she thinks of her body size/shape
- binge eating
- vomiting
- laxatives
- exercise
- mood/ sleep pattern/ anhedonia
- previous psychiatric illnesses
- menstrual history
- general medical history
- medication/ drugs/ smoking/ alcohol

Next steps

- discusses likely diagnosis of anorexia nervosa, without antagonising the patient.
- suggests blood tests to check thyroid, U&Es (especially K+), FBC (anaemia)
- agrees a plan for review
- may discuss treatment with talking therapies as the primary modality

Appropriate questioning technique (mixture of open and closed questions)

Avoids or explains jargon

Uses tools such as signposting and summarising

Systematic, logical approach

Checks if the patient has any other questions

Finishes consultation appropriately

Friendly approach and appropriate body language

68 Notes

Note: if the candidate asks for the patient's weight, tell them that after taking the height and weight of the patient, the BMI is found to be just under 16.

69 History   Candidate role

You are a Foundation doctor working in casualty.

A confused young man has been brought in by a friend. Please take a history from the friend.

After your history be prepared to describe

- your differential diagnoses
- the next steps you would take, with justification for why you think they are important.

69 History   Patient role

You are a friend of John Brown, aged 25.

Only give the following information if asked directly:

You went to visit him today at his flat. He was acting very strangely. At first he would not let you in the flat at all. He was shouting bizarrely – talking about the gangsters coming to get him. He seemed to be staring wildly and looked really odd. It is difficult to describe exactly what was wrong – he kept talking rubbish and seemed to be frightened of something but you could not work out what. He had piled books and boxes up against one of the windows of his flat – he said it was a defence against spies. He looked like he had not had a wash for a few days - his flat smelt quite bad.

You are really worried. You did not know what to do. You do not know his family at all so could not call anyone. It took you a while to persuade him to come with you to the hospital at all.

Background

You have known John a year, since you both arrived at university. You think his family live in Kenya – his dad is a diplomat or something like that.

You do not know if John has been in hospital before. He uses an asthma inhaler but you do not know if he takes any other medicines.

He drinks – like all students! – maybe 5 or 6 beers once or twice a week as far as you know. He has smoked cannabis occasionally when you have been out – but doesn't everyone. You also know he once used E in a club, but you have never known him take anything stronger.

You last saw him a week ago when you had a night out clubbing to celebrate being back at Uni after the summer break. He seemed OK then. He did have a bit of a fight with a bloke in the club, it seems like he had recently split up with his girlfriend and this bloke was chatting her up in the club. He was pushed around a bit, but he gave as good as he got. You left shortly after the fight and you think John probably went home a bit later.

69 History   Examiner mark sheet

Introduces self and role and checks identity of friend
Identified the speaker and their relationship to the patient
Established events immediately prior to admission
- how was patient found?
- any clues as to what had happened?
- when was he last seen beforehand (behaving normally)?
- other witnesses who might know more?

Established how patient is normally (i.e. level of functioning normally)
Past medical history
Drug history (prescribed and recreational)
Allergies
Social history
- who is at home
- occupation
- alcohol
- smoking

Asks directly about anything else relevant not already mentioned:
- head injury?
- bottles of pills?
- recent travel?

Appropriate questioning technique (mixture of open and closed questions)
Avoids or explains jargon
Uses tools such as signposting and summarising
Systematic, logical approach
Checks if the patient has any other questions
Finishes consultation appropriately
Friendly approach and appropriate body language

69 Notes

Differential should include:

- drugs (recreational)
- drugs (prescribed possibly in overdose)
- head injury (fight)
- tropical illness (from Kenya)
- psychosis
- physical illness (unknown but to include diabetes, hyper/hypocalcaemia, infection including encephalitis)

The next steps would be to
- examine the patient looking for injuries/ smells/ signs of drug abuse/ signs of particular drugs e.g. pinpoint pupils/ chest infection/ meningism...
- investigations including drug screen, FBC, U&Es, Glucose, Calcium, CRP,
- if there are signs of injury, X-ray the injured part, if the chest has signs CXR, consider head scan (MRI/CT) after discussion with the radiologist

Miscellaneous practice scenarios

70 History   **Candidate role**

You are a junior doctor covering the psycho-geriatric ward in a split site hospital.

A nurse from the other site has asked you to call her back regarding Ann Brown, aged 67. According to the message left the nurses have noticed that Mrs Brown is itching – scratching her body.

Mrs Brown suffers from Alzheimer's' dementia so cannot give you any history herself.

- Please telephone the nurse in charge and take a brief history to enable you to make a differential diagnosis until you can get to the ward.
- Ask the nurse to initiate investigations, if you feel they are required.
- At the end of the conversation, tell the examiner the differential diagnoses you are thinking of.

70 History   Patient role

You are the nurse in charge of the psycho-geriatric ward in a split site hospital in Manchester. Mrs Ann Brown, aged 67, (one of your patients with Alzheimer's' dementia) has started scratching herself. She has been doing this for a few days now. You notice she is always itching when she is in the social area, but you decided to call the doctor when you bathed her and found that she is covered with scratch marks. You left a message and now the doctor is calling you back.

Please answer the doctors questions but do not offer information that is not asked for directly:

Ann has no rash, other than the scratch marks where she has been itching. These are mostly on her back but also on her arms and legs – really anywhere she has been scratching. She has never had this problem before.

Background

She seems slightly more agitated than is normal for her and has had the hiccups a few times recently – but you thought nothing of that. Otherwise nothing much has changed. She always has looked pale. She can express when she is distressed but not much else that really makes sense. She can feed herself but has to wear incontinence pads as she tends to be incontinent of urine.

No-one else in the ward is scratching (you have had scabies on the ward before).

Ann is a diabetic on tablets with well controlled high blood pressure. She is on quite a lot of tablets:

- Metformin 500mg tds
- Ramipril 5mg od
- Aspirin 75mg od
- Atorvastatin 40mg od
- Bendrofluazide 2.5 mg od
- Reminyl (galantamine) 8mg bd (started four weeks ago, building up the dose)
- Paracetamol 500mg 2 QDS

You do not routinely take blood sugars on Ann as she is not on insulin.
You are able to take blood from Ann if the doctor asks you to do some initial investigations before he can get to see her.

70 History   Examiner mark sheet

Introduces themself in a professional manner
Establishes identity of the person answering the phone and of the patient
Elicits overview and details of the problem
Clarifies/checks relevant main points of what has happened
Specifically asks about:
- Site of itching
- Is the itching generalised? Or localised as a rash?
- Any infectious contacts (others in ward itching)?
- Previous history of itching?

Asks about past medical history – specifically:
- diabetes
- thyroid
- anaemia
- lymphoma
- kidney disease
- liver disease

Asks about drug history (including recent changes in drugs)
Agrees plan for further actions:
- Urea & Electrolytes (kidney function)
- Liver Function Tests
- Thyroid Function Tests
- Full Blood Count
- Ferritin
- HbA1C and BM stick glucose
- CRP/ESR

Provides opportunities for questions
Checks understanding of information given
Summarises main points of discussion and decisions
Easily understood, with appropriate language
Polite, concerned, friendly manner throughout the conversation

70 Notes

Differential diagnoses of generalised itching

- High blood sugar
- Contact dermatitis (soaps, washing powders, woollen clothing)
- Drug reaction (as recently started new medication)
- Renal failure (important to rule out especially in view of hiccups)
- Scabies (there is always a first patient to be infected)
- Thyroid problems
- Anaemia/ low iron/ polycythaemia
- Lymphoma
- Liver disease

71 History/ Explanation   Candidate role

You are a Foundation doctor in General Practice, tackling the advice phone calls. The receptionists have noted down the following message:

    Jo Johnson
    23 years old
    Spots on stomach, bit unwell 2/7
    Wants advice.

Please phone the patient back and advise as appropriate.

71 History/ Explanation   **Patient role**

You are a 23 year old primary school teacher (newly qualified).
You are previously fit and well and on no medication.
You live on your own.
This morning you have woken up with spots on your stomach. You have been feeling a bit under the weather for a couple of days, but thought it was the stress of the new job. Some of them are just pink bumps but some of them are like little (half centimetre diameter) blisters. They itch a bit and one or two have scabs on from where you scratched them in the night. Funnily enough you feel better today than you did yesterday – that achy joints feeling has settled down.

If asked what you think it might be, tell the doctor:
You know that there is a lot of chickenpox going around the school. You suspect that you have chicken pox and want advice as to what you can do for it.
You had heard that calamine lotion is good for chickenpox.
How long should you be off school?
And can the doctor give you a sick note?

71 History/ Explanation   **Examiner mark sheet**

Introduces self and role
Checks identity of caller
Explains/ confirms purpose of interview
Establishes nature of presenting complaint
Description of rash – site (distribution), onset, nature of spots,
Associated symptoms - fever, neck stiffness, cough or breathlessness, confusion
Establishes how patient has self-treated already
Drug history
Past medical history
Explores patient's ideas, concerns, expectations

Makes diagnosis of chickenpox
Advises simple remedies –
  • Calamine lotion
  • lukewarm baths (heat makes itching worse)
  • antihistamines
Advises to stay away from pregnant women and people who are immunosuppressed
Off school until the spots are all scabbed (about one week)
Doctor can leave a sick note without seeing patient in the flesh
Safety nets – offers review if worried/ worse

Appropriate questioning technique (mixture of open and closed questions)
Avoids or explains jargon
Uses tools such as signposting and summarising
Systematic, logical approach
Checks if the patient has any other questions
Finishes consultation appropriately
Friendly approach, fluent professional manner throughout

71 Notes

72 Explanation   Candidate role

You are a Foundation doctor on the gastroenterology wards.

Your next patient (aged 25) is just about to be discharged home. The patient has been diagnosed with ulcerative colitis and started on high dose steroids (prednisolone 40mg daily). The plan is to maintain this dose until review in out-patients, then the dose may be reduced.

You have been asked to discuss the medication with him/ her before discharge.

You do not have to take a history or explain the diagnosis.

Please ensure the patient has adequate information about taking high dose steroids.

72 Explanation   Patient role

You are 25 years old.

You came into hospital after a month of bloody diarrhoea. You were feeling pretty awful but are recovering now. You have been started on steroid tablets and understand that you are to take these at home after discharge.

The Foundation doctor is going to come to talk to you about the tablets before you go home.

You may have a few preconceptions about steroids. Does this mean that you will be an Olympic athlete now?!! If you are a woman, you may be have memories of female athletes who took steroids in the past and grew facial hair and big Adam's apples!

How long will you have to take the tablets?

And why don't they make them in bigger doses. It is a nuisance taking 8 small tablets at the same time. Can't they give you one 40mg tablet instead?

Background

You are thinking of planning a family in the next six months, will that be OK?

You work in an insurance office. You drink a few glasses of wine/ beer at the weekends.

You don't smoke.

You are not on any other medications and have never been in hospital before.

72 Explanation   **Examiner mark sheet**

Introduces self and identifies role

Checks patient's identity

Confirms/ establishes reason for talk today

Establishes understanding of treatment

Facts to include in explanation:

- Steroids must not be stopped suddenly (as this can cause collapse & life-threatening illness)
- Patient must carry a steroid card so that other doctors will know they are on steroids if an emergency occurs.
- If the patient is ill they may need a higher dose of prednisolone, so must see a doctor quickly.
- When the consultant decides to reduce the dose, or stop the steroids, this will be done slowly over several weeks

Side effects:

- weight gain (due to increased appetite)
- osteoporosis (thinning of the bones) if taken long term. This can be prevented by medication which will be prescribed if needed.
- indigestion (again can be treated)

Establishes patient's concerns, validates and addresses these (effects of steroids; number of tablets; thinking of starting a family)

Uses chunking and checking - giving small pieces of information and checking understanding before continuing

Checks if patient has other questions

Uses communication tools, such as signposting and summarising, appropriately

Offers written advice or websites

Language appropriate throughout

Friendly approach, appropriate body language

Fluent and professional manner

72 Notes

LEARNING POINTS

Prednisolone is usually prescribed as 5mg tablets in the UK. This often means the patient must take a large number of tablets, but it allows for easy dose adjustment.

The type of steroids used (prednisolone) will not cause masculinisation or athletic ability.

Steroids have many potential side effects which you must know about, but you would not mentions them all at the first discussion with the patient. These include increased blood sugar; increased risk of cataracts; increased risk of infection

When discussing medication with women of reproductive age, always consider the need for contraception and whether the tablets are safe in pregnancy. In this case, it would be sensible to suggest waiting to start a family until disease has settled and ensure the patient talks to consultant about this.

73 History/explanation   Candidate role

You are a Foundation doctor in General Practice. Your next patient is originally from South Africa. Last week the patient asked the practice nurse to take blood for an HIV test, saying that the test was needed for a life insurance application. The nurse was unsure about HIV testing and advised the patient to see the doctor.

Please counsel your patient with regard to having the HIV test.

73 History/ Explanation

Patient role

You are 37 years old. You have applied for life insurance and the firm has asked you to have an HIV test. You came to the practice nurse and she suggested you see the doctor, who is now going to see you. As far as you are concerned right now it is just a simple blood test and you don't think it will be positive.

Background

You are originally from South Africa, but grew up from the age of three in the UK. You have travelled back to South Africa for family holidays, but otherwise your life is in Britain. Your partner is Nigerian –born and brought up in Africa but has lived in the UK for the last five years. You met in a nightclub. You know that you both have had previous relationships but you have not shared details. You used condoms at first, but not once the relationship was established – you can trust each other and prefer sex without a condom, it is more spontaneous. You do not have anal sex.

You have had previous sexual partners – but not in the last couple of years. You have never had any homosexual relationships and are horrified at the idea. You would be insulted if you thought someone was suggesting that you might be homosexual.

You smoke occasional marijuana but have never injected drugs.

You have a tattoo which you had done during a holiday in South Africa – it represents traditional tribal markings.

You had an operation when you were a child (in the UK) – you had your tonsils out. You cannot remember if you had a blood transfusion at the time.

You are normally fit and well. You smoke a few cigarettes when you are out with friends and drink beer – maybe three times per week a few lagers. You are on a tablet for blood pressure but cannot remember the name of it.

73 History/ Explanation   **Examiner mark sheet**

Introduces self and identifies role
Checks patient's identity
Confirms/ establishes reason for visit
Confirms confidentiality of discussion
Establishes the patients concerns
- why test now?
- previous tests?
- understanding of and exposure to risk factors?

Establishes patient's understanding of risks and corrects any misunderstandings:
- men who have sex with men
- IV drug use/sharing needles
- blood transfusions overseas
- tattoos/body piercing overseas
- sex with people from areas with a high prevalence of HIV such as sub-Saharan Africa

Discusses false negatives due to "window" of infection – need to retest in three months if recent risk behaviour
Discusses health reasons for having test:
- treatment is available for HIV that controls the disease and allows a full, healthy life
- protect partners/ family if positive

Discuss when and where results will be given
Negative result does not mean risk behaviour is OK
Offers written information
Asks the patient if they have any questions
Friendly approach, appropriate body language
Fluent, professional manner throughout

73 Notes

> **LEARNING POINT**
>
> Late diagnosis is one of the most significant risk factors for poor prognosis. Diagnosed early HIV is now considered to be a chronic treatable disease which patients can live with for many decades.
>
> Insurance companies are not allowed to discriminate against people simply because they take a test, but refusing to take a test if requested to by an insurance company might lead to much higher premiums or refusal of insurance.
>
> Under current British HIV Association guidance, testing for HIV should become much more routine in the UK health service than it has been in the past. Hospitals and GPs are recommended to offer testing to all who may be considered to be at risk. This includes all new patients coming from populations where HIV prevalence is greater than 2/1000, such as areas of London and big cities such as Manchester.
>
> See
>
> **www.bhiva.org/documents/Guidelines/Testing/GlinesHIVTest08.pdf**
>
> and be sure to find out the local policy regarding HIV testing in your organisation.

74 History/explanation   Candidate role

You are a Foundation doctor on the emergency medical admissions unit. The phlebotomist comes to you very upset. Whilst taking blood from a patient, she gave herself a needle stick injury. She is worried about the possibility of having contracted a viral infection – hepatitis or HIV.

The occupational health policy requires that you counsel the patient and request consent for taking blood to check for infection. Please discuss this with the **patient**.

NB: The patient here is not the phlebotomist. It is the patient from whom blood had been taken when the phlebotomist gave herself a needle stick injury.

74 History/ Explanation   **Patient role**

You are 29 years old and are in hospital because you fell off your motorbike and broke your leg. You have had surgery to put a plate in your leg, and are hoping to get home soon. The nurse took blood this morning, you are not sure why, but you think that if the test was Ok you would be likely to get home tomorrow.

The doctor is coming to speak to you now, but you do not know what for. You are likely to get pretty annoyed if it is something that will delay you going home, or cause you more stress. You just don't need it right now.

Background

You work as a sound engineer for a variety of rock bands. It is not a bad job, at least it keeps you out of the nine to five routine. You spend several months of the year touring with the bands. But now you will not be able to work for months and you are quite worried about that.

You live with your partner, Mike.

You drink a few beers three or four nights a week, and smoke an occasional spliff. You have tried E in the past but nothing stronger. Not sure about Mike though – he has a bit of a wild past. Well everyone has their history don't they? You have lived a bit with being on the road so much...

You have a couple of tattoos, a dragon on your shoulder that you had done in Thailand a couple of years ago, and a traditional good luck symbol at the bottom of your back.

The only other time you have been in hospital was to have your appendix out when you were fourteen. You have no idea if you had a blood transfusion, and you do not know your blood group.

74 History/ Explanation   **Examiner mark sheet**

Introduces self and identifies role

Checks patient's identity

Explains about needle stick injury to health worker
- no risk to patient
- very small risk to health worker
- better if patient can be tested

Establishes what the patient's concerns, validates and addresses these

Establishes patient's understanding of and exposure to HIV and hepatitis risk factors
- men who have sex with men
- IV drug use/sharing needles
- blood transfusions overseas
- tattoos/ body piercing
- sex with people from sub-Saharan Africa

Discusses benefits of testing
- treatment available to cure hepatitis and to control HIV
- knowledge allows patient to protect partners and family

Discusses false negatives due to "window" of infection – need to retest in three months if recent risk behaviour

Discusses HIV testing and insurance cases (a negative test does not count against patient – refusing a test might)

Discuss behaviour after test – avoiding risk behaviour

Describes when and where results will be given

Uses chunking and checking - giving small pieces of information and checking understanding before continuing

Checks if patient has other questions

Uses communication tools, such as signposting and summarising, appropriately

Offers written advice or websites

Language appropriate throughout

Fluent and professional manner

Friendly approach with appropriate body language

74 Notes

75 History   Candidate role

You are a Foundation doctor in General Practice.

Your next patient is a 41 year old man who has contacted the surgery by telephone, complaining of a painful lump in the left side of his groin.

Please speak to him and suggest an appropriate course of action.

75 History   Patient role

You are a 41 year old man. You have telephoned the doctor this morning because you don't think you can get to surgery. You have a big lump in your groin. It is real agony.

You live on your own, and you doubt you could walk or drive to the GP, you don't want to risk making it worse.

You really want the doctor to come out. You are quite frightened by this and in a lot of pain.

Background

You noticed a slight lump in the left side of your groin a couple of week ago, but at first it was not painful, just a bit heavy and dragging. It seemed to disappear when you had been to sleep, so you did not think much about it. This morning you find that the lump is suddenly much bigger, and getting more and more painful. It doesn't go away if you press on it or lie down, like it used to.

The lump is about as big as your fist – it is right in the angle between your left leg and your nuts – and it hurts if you press on it. It is not red but it does feel a bit warm to touch. You do not remember a specific injury that has caused it.

The pain does not seem to be relieved by anything – you only have paracetamol in the house and have taken two about half an hour ago

You are feeling sick through pain but have not actually vomited – you did not feel like any breakfast this morning so you have only had a sip of tea to drink

You moved your bowels as usual yesterday but have not been yet today

You work as a labourer on construction sites

You have no other illnesses and do not normally take medicines.

There is no-one who could run you to the surgery and you cannot afford a taxi.

75 History

Examiner mark sheet

Introduces self
Confirms who they are speaking to
Finds out the reason for the call
Takes a brief history of lump:
- location;
- size;
- onset;
- duration/coming and going or always present;
- exacerbating and relieving factors;
- severity of pain;
- previous occurrence

Establishes past medical history
Establishes drug history
Establishes concerns, addresses and validates these (patient feels he needs a home visit)

Recognises the presentation as an emergency and advises the patient that he needs to be seen urgently by a doctor – either the GP (on a home visit if necessary, or A&E)
Safety nets by
- Making sure the patient is able to get to a doctor (either offer a home visit or get him to the surgery or hospital – suggest a family member, a neighbour or a taxi or if not possible an ambulance)
- Checking that the patient has understood the seriousness of the situation
- Ensuring patient will re-contact if problems with the planned management

Appropriate questioning technique (mixture of open and closed questions)
Avoids or explains jargon
Uses tools such as signposting and summarising
Systematic, logical approach
Checks if the patient has any other questions
Finishes consultation appropriately
Friendly approach and appropriate body language

75 Notes

> LEARNING POINT
>
> If you think that the patient should be seen as an emergency, make sure that they understand the time frame in which you want them to be seen and will call again for help if plans fall apart. GPs have been caught out in the past when unexpected delays have occurred and patients did not realise the need to call back.

76 Explanation   Candidate role

You are a Foundation doctor in A&E.

The next person you have to deal with is the son/ daughter of a 71 year old man, who had surgery one week ago in the same hospital, for repair of left inguinal hernia.

The patient has just been seen with a wound infection that may need re-admission. He has given permission for his relative to speak to you.

Please deal with the family's concerns.

76 Explanation   Patient role

You are the relative of a patient in casualty. Your father has given permission for you to speak with the doctor and now you want a word...

Your father is 71 years old. A week ago he had a hernia operated on. Now the wound is red, swollen and mattering (it's leaking pus - disgusting) and very painful for your Dad.

You are angry because your Dad is an old man. It was bad enough that he had to wait so long for the operation. Now this...

These bloody doctors don't care, they get paid lots of money and have fancy cars and big houses. Your friend works in a hospital and she said that it was probably the surgeon being careless. You've seen reports in the papers about patients getting infections in this hospital. Your Dad thinks it was one of the trainee doctors who did the operation – it must be that doctor's fault. They probably hadn't been properly trained yet.

Your Dad worked hard all his life, and now this. You are really quite angry about this and want to get sorted out quickly and make a complaint against the doctor who did the surgery.

76 Explanation   Examiner mark sheet

Introduces self and identifies role

Checks patient's and relative's identity

Confirms patient has given permission to talk (possibly by including patient in discussion)

Confirms/establishes reason for visit

- establishes concerns (wound infection),
- acknowledges the distress caused
- apologises for the circumstances
- Explains that wound infection sometimes happens even when the proper procedures had been followed.

Finds out what relative would like to happen next

Offers positive actions

- Prompt treatment of the infection to get the patient better as fast as possible
- If the relative still wants to make a complaint, suggest a suitable course of action for this to happen (calling a senior doctor to speak to the patient; or explaining the hospital complaints procedure; or referring the relative to PALS, the patient advocacy and liaison service)

Reaches agreement about further management

Checks for unanswered questions

Remains calm and polite throughout

Language appropriate throughout (avoids jargon and avoids inflammatory language)

Fluent and professional manner

Friendly approach, appropriate body language

Checks if patient has other questions

76 Notes

> LEARNING POINT
> This station tests your ability to deal with an angry patient or relative. It also illustrates why the consent procedure is so important. We must always mention(and record mentioning) the possibility of wound infection during consenting so that the patient will be more prepared should this happen – even though we try to avoid infection where ever possible.

77 Consent   Candidate role

You are a Foundation doctor in a surgical ward.

Your next patient is a 53 year old who has had laparotomy for oversewing of a bleeding duodenal ulcer. Post operatively s/he has a paralytic ileus and requires a naso-gastric tube to assist drainage of the stomach and decompression of the bowel.

Please explain the procedure of inserting the naso-gastric tube.

77 Consent

Patient role

You are 53 years old.

You are on the surgical ward, having had an operation on your bowel for a bleeding ulcer yesterday evening. Now you are getting stomach pain and vomiting small amounts of green bile. It is very unpleasant. Your stomach feels all blown up with gas. The doctor has come to explain what they will do to help this.

You are worried that there was a problem with the operation and want to ask the doctor about this.

77 Consent   **Examiner mark sheet**

Introduces self and identifies role
Checks patient's identity
Confirms/ establishes reason for talk
Establishes understanding of the current situation
Facts to include in the explanation:
- The bowel has gone to sleep after the operation, so the gastric juices and gases build up
- A nasogastric tube will help relieve the feeling of nausea and help prevent the patient from vomiting by keeping the stomach empty

What will happen (before, during and after the procedure)
- A thin tube is inserted through the nose, down the back of the throat and into the stomach
- The tube is about the same thickness as an electrical cord
- The patient will need to swallow as the tube goes down
- Once the tube is inserted, an X-ray will be taken to ensure it is in the right place
- The tube will remain in place for a day or so to rest the bowel and until the bowel starts to work again

Minor common complications
- It can be a little uncomfortable when the tube is inserted, but is not usually uncomfortable when the tube is in place
- Some patients have a slight sore throat for a few days after having a tube in place

Serious rare complications
- There is a small chance that the tube might enter the airway instead of the gullet. This would be very uncomfortable, make the patient cough, and the doctor would immediately take the tube out

Uses chunking and checking - giving small pieces of information and checking understanding before continuing
Establishes patient's concerns, validates and addresses these (it does not mean the operation went wrong, this sometimes happens to some patients after a procedure like the one she had)
Checks if patient has other questions
Uses communication tools, such as signposting and summarising, appropriately
Language appropriate throughout
Friendly approach, appropriate body language
Fluent and professional manner

77 Notes

LEARNING POINT

Nurses may be trained to insert NG tubes, but it is useful for junior doctors to become skilled in this too, as they may be asked to insert a tube if the nurses are busy or having difficulty with it.

Nasogastric tubes can be used for draining the stomach, as here, or in other situations for feeding a patient who temporarily cannot eat. It is important after inserting a tube to establish that it is in the right place, especially if feeding via the tube is to be contemplated. Two methods of checking this are recognised: X-ray and checking of the pH of withdrawn fluids (stomach content is acid). In the past it was practice to squeeze a syringe full of air down the tube and listen with a stethoscope for the sounds of gas in the stomach. This is no longer accepted as safe practice as several patients have died of aspiration pneumonia as a result of being fed down misplaced tubes.

78 History   Candidate role

You are working on the wards. A nurse calls you to say that a patient has become confused during the course of the evening. The patient himself cannot give you any useful history. Please take a history from the nurse.
After your history describe
- your differential diagnoses
- the next steps you would take, with justification for why you think they are important.

78 History   Patient role

You are the nurse on duty in the surgical ward. You have just been doing the observations for the night and found one of your patients, Mr Davies, has become confused.

Only give the following information if asked directly:
Mr Davies is an elderly (83 year old) man who had an ERCP earlier today.
He was only admitted this morning. Normally the patients for ERCP are done as a day case but you understand he lives on his own and so it was felt that he should be kept in overnight. As far as you know the procedure went OK. You heard he had a stone removed from his bile duct.
He seemed a very sweet old man when he was admitted but now he is shouting and causing a rumpus in the ward. He is seeing gypsies and swearing at them.

You can't find his notes in the ward but do have his medication chart.
On his medication chart he has
Co-codamol 8/500 2 qds prn
Bendrofluazide 2.5mg od
Madopar dispersible tablets 62.5mg tds

Give only the observations requested:
Temp 37.5
P 104
BP 105/65

Oxygen saturation 91% on air

You do not know if he has passed urine today – he is not on a chart.
He is not complaining of any pain – but then he is not making much sense at all.

78 History   Examiner mark sheet

Introduces self and role and checks identity of patient and speaker

Explains/confirms purpose of interview

Establishes patient's history as far as nurse is aware

- today's procedure
- how patient was at admission
- past medical history as far as known
- drug history
- symptoms that might provide clues
    - pyrexia
    - coughing
    - pain
    - PU

Asks for vital signs P, BP, temperature, O2 saturation, fluid balance

Appropriate questioning technique (mixture of open and closed questions)

Uses tools such as signposting and summarising

Systematic, logical approach

Checks if the nurse has any other questions or information

Finishes appropriately

Friendly approach and appropriate body language

78 Notes

Differential diagnoses

- infection (cholangitis most likely, also exclude UTI and chest infection)
- pancreatitis or perforation from ERCP
- drugs (prescribed e.g. madopar)
- electrolyte imbalance (e.g. from anti-hypertensives)
- cardiac event
- alcohol withdrawal due to hospital admission
- increased confusion in mildly dementing patient due to change of environment

The next steps would be to

- ask the nurse to put oxygen on the patient (high flow)
- examine the patient looking for signs of infection/ signs of abdominal pathology
- bloods including glucose, U&Es, FBC, calcium, amylase
- MSU
- ECG
- an erect CXR
- Call for senior advice

LEARNING POINT

Reviewing a patient post-procedure would be an everyday responsibility for a junior doctor but perhaps not one that many students have considered or prepared for. Students tend to focus on the "new" presentation, rather than a deterioration of a patient already on the ward.. It is important that you are confident in initial recognition and management of conditions such as new onset of confusion; chest pain; breathlessness; abdominal pain; pyrexia in post-operative patients.

---

79 Consent   Candidate role

You are a foundation doctor on the surgical ward.

Your next patient has been in your ward with his smelly, ulcerated and gangrenous foot for the past couple of weeks. The X-ray shows that he has osteomyelitis. Conservative management of his ulcers has failed. He is in pain and the leg smells badly.

Your consultant has suggested that amputation is really the best option now. It would relieve the pain and remove the smell. And the patient could be rehabilitated on a prosthetic leg, or given a wheelchair if appropriate.

Please consent the patient.

79 Consent

Patient role

You are aged 68 years

You have been diabetic for 15 years now – you are taking insulin injections as well as tablets. Your left leg has had ulcers for five years now. They are quite painful and sometimes smell bad.

Life is quite tough.

You are a retired bus driver.

You live alone in a second floor flat. Your other half died three years ago.

You don't like to go out much other than the pub where you are known, because you have to use sticks and walking is difficult. There is no lift in your flats so you have to get down the stairs. It would be impossible in a wheelchair like some cripple.

Other than the pub, you enjoy looking after your birds – you have two parakeets. Your friend is looking after them whilst you are in hospital.

The doctors have suggested previously that your leg is not getting any better (as if you need them to tell you this!). They think you should have it cut off. You would rather have the pain and the smell than do this. A mate lost a leg in a road traffic accident. Afterwards his life was hell. Eventually he turned his face to the wall and died. You are not prepared to contemplate amputation no matter what the doctors say.

79 Consent

Examiner mark sheet

Introduces self and identifies role

Checks patient's identity

Confirms/ establishes reason for visit

Elicits patient's ideas and concerns, validates and addresses these (friend's history; not getting up and down stairs in flat; looking after birds)

Discusses benefits of amputation and how patient's concerns could be addressed:

- Pain (and smell) would be much less
- Could be rehabilitated in a wheelchair or with a prosthesis
- Could assist in getting ground floor flat

Checks patient has capacity

- Screens for depression
- Checks understanding of information
- Checks retention of information
- Ensures patient understand consequences of not operating

Reassures patient of support and care whatever the decision

Reassures patient that decision is not now or never

Offers other sources of information (visit from OT and physiotherapist; written leaflets or websites)

Checks if patient has other questions

Uses chunking and checking - giving small pieces of information and checking understanding before continuing

Uses communication tools, such as signposting and summarising, appropriately

Language appropriate throughout

Fluent and professional manner

Friendly approach with appropriate body language

79 Notes

LEARNING POINTS
This station should prompt you to revise the mental capacity act.
http://www.justice.gov.uk/guidance/protecting-the-vulnerable/mental-capacity-act/index.htm
For a patient to have capacity they must be able to
Understand information
Retain it
Analyse it to make a decision
Communicate that decision to you
And their decision should not be unduly influenced by the presence of a mental illness that clouds their judgement.

The five principles of the mental capacity act are:

You must assume an adult has the capacity to make decisions about the future unless it is proven otherwise.

You must give all practicable help to enable patients to understand decisions before deciding that they lack capacity. This may mean giving information in different languages, in simplified versions, with pictures or with dolls if necessary.

Just because an individual makes what you see as an unwise decision, does not mean they lack capacity. You may not share their beliefs of lifestyle, so what to you seems unreasonable to them may be entirely rationale. Ask about their reasons and see if there is anything amenable to help.

Any decision made on behalf of a person who lacks capacity must be done in their best interests.

Anything done for or on behalf of a person who lacks capacity should be the least restrictive of their basic rights and freedoms. This means you should not choose an irreversible option for a person lacking capacity, where a reversible option is likely to give a similar result. Thus reversible contraception might be preferred over sterilisation for example.

80 Complaint   Candidate role

You are a Foundation doctor in orthopaedic surgery.

You are asked to speak to a relative who is visiting a 68 year old patient on your ward. The patient had a knee replacement (due to osteoarthritis of the joint) a week ago. A couple of days after the operation, he developed a fever and his wound site appeared infected. He has been found to have MRSA in his wound and is now on appropriate antibiotic treatment.

Please speak to the relative and deal with the concerns raised.

80 Complaint   Patient role

You are visiting your father, who is 68. He is a patient on an orthopaedic ward as he had a total knee replacement a week ago.

You have come to visit him today and he has told you that he has an infection in his wound which the doctors have told him is MRSA. You are quite shocked by this – you have read about MRSA and it is one of these superbugs that is very difficult to treat. It has also meant that he hasn't been able to do the physiotherapy he should be doing.

You have travelled from London to be with your father after his operation (he is a widower). You ask to speak to one of the doctors to find out what is happening – you want to know why it happened and are quite angry that it has happened. Maybe the surgeon was careless?

80 Complaint

Examiner mark sheet

Introduces self and identifies role

Confirms who they are speaking to - identity and relationship to patient

Confirms patient consent to talk with relatives (possibly by including patient in discussion)

Confirms / establishes reason for discussion

Establishes concerns, validates and addresses these (worry that infection may be very difficult to treat and anxious over delay to recovery)

Facts to include in the explanation:

- acknowledges the distress caused
- apologises for the circumstances
- emphasises condition was detected and is being appropriately treated
- discusses implications:
    o The patient is likely to be in hospital longer than otherwise would be the case (possibly weeks)
    o He may need to be nursed in a single room with special anti-infection precautions.
- discusses infection as one of risks of operations, which is explained to all patients
- the hospital does look into all cases of MRSA to try to learn what if anything more could be done to prevent such infections

Manages to calm fears and anger of relative successfully

If unable to calm relative offers senior to talk (or complaints procedure)

Checks for unanswered questions

Remains calm and polite throughout

Language appropriate (avoids or explains jargon)

Fluent and professional manner

Friendly approach with appropriate body language

80 Notes

> LEARNING POINT
>
> Hospital acquired infection is a big problem in modern hospitals, partly because the patient population is aging and becoming more vulnerable, partly because use of antibiotics has led to more resistant organisms.
>
> You need to be fully aware of the appropriate hygiene precautions, both those used for all patients and those used for patients known to be infected with a resistant or highly contagious organism.
>
> Familiarise yourself with the issues around avoiding, detecting and treating MRSA and *C. difficile*.

Appendix 1 - generic mark sheets

History   **Examiner Mark sheet**

Introduces self and role and checks identity of patient
Explains/confirms purpose of interview
Establishes nature of presenting complaint (as relevant)
- Site
- onset
- character
- radiation
- time duration
- exacerbating and relieving factors
- severity

Enquires about relevant associated symptoms:
- …

Establishes effect on daily life

Explores ideas, concerns, expectations

Drug history (including over the counter and illicit drugs)

Allergies
Establishes previous medical history

Establishes family medical history

Establishes social history
- occupation
- who is at home
- smoking
- alcohol

Excludes other systemic symptoms

Appropriate questioning technique (mixture of open and closed questions)
Avoids or explains jargon
Uses tools such as signposting and summarising
Systematic, logical approach
Checks if the patient has any other questions
Finishes consultation appropriately
Friendly approach and appropriate body language

Explanation   **Examiner mark sheet**

Introduces self and identifies role
Checks patient's identity
Confirms/ establishes reason for visit
Establishes understanding of test and of the expected result
Facts to include in the explanation:
    What the result is

    What that means/ why it is a problem

    What can be done about it/ how to prevent complications

    Follow up

Establishes patient's concerns, validates and addresses these
Uses chunking and checking - giving small pieces of information and checking understanding before continuing
Checks if patient has other questions
Uses communication tools, such as signposting and summarising, appropriately
Offers written advice or websites
Language appropriate throughout (avoids jargon)
Fluent and professional manner
Friendly approach, appropriate body language

Consent   **Examiner mark sheet**

Introduces self and identifies role
Checks patient's identity
Confirms/establishes reason for visit
Establishes understanding of test and of expected/ possible results
Facts to include in the explanation:

- What will happen and why (before, during and after the procedure)

- Minor, common complications

- Rare, serious complications

Establishes patient's concerns, validates and addresses these
Uses chunking and checking - giving small pieces of information and checking understanding before continuing
Checks for unanswered questions
Uses communication tools, such as signposting and summarising, appropriately
Language appropriate throughout
Fluent and professional manner

Appendix 2 – crib sheets for history and examination

The crib sheets that follow can be photocopied and used to assist you in taking your histories and recording examinations. With time you should commit the routine to memory, and learn to select the most relevant questions as time allows.

FULL CLERKING SHEET - History

Presenting complaint – P/C or C/O

History of presenting complaint

Effect on daily life – consider work, home (family, relationships), social (hobbies, friendships), financial life
Ideas
Concerns
Expectations

Drugs
- Prescribed
- Non-prescribed – from the chemist and recreational
- Allergies

Past medical history, especially:
- Operations or other hospital admissions
- Diabetes, asthma, high blood pressure, jaundice, epilepsy

Family history

Social history
- Occupation
- Who is at home?
- Smoking
- Alcohol
- Pets?
- Travel?

Systems enquiry
Cardiovascular system
- Chest pain
- Palpitations
- Breathlessness – SOBOE, PND, Orthopnoea
- Ankle oedema
- Claudication

Respiratory system
- Cough
- Breathlessness
- Wheeze
- Spit
- Haemoptysis

Gastrointestinal system
- Appetite – nausea, vomiting, haematemesis
- Weight
- Bowel habit – recent change, constipation, diarrhoea
- PR bleed
- Indigestion
- Abdominal pain

Genitourinary system
- Dysuria
- Haematuria
- Continence
- Prostatism – flow, hesitancy, terminal dribble, nocturia

- LMP
- Cycle
- Post coital or intermenstrual bleeding
- Vaginal discharge
- Contraception

Central nervous system
- Fits, faints, funny turns, black outs
- Headaches
- Dizziness
- Numbness, tingling, weakness
- Eyesight
- Hearing

Other
- Musculoskeletal – joint pains or swelling
- Rashes, lumps or bumps

Examination
- General observations
- Pallor/cyanosis/jaundice
- Nodes/clubbing/goitre

CVS
- Pulse
- BP
- JVP
- HS   Apex beat
- Murmurs/bruits   Thrills

Respiratory
- Trachea
- Expansion
- Percussion note
- Breath sounds
- Added sounds

Abdomen
- Inspection
- Palpation - masses
  - organomegaly
  - tenderness, rebound, guarding
- Percussion
- Bowel sounds

CNS
- Mental state – oriented in time place and person?
- Motor
  - Power
  - Tone
  - Reflexes
- Sensory
  - Light touch
  - Pin prick
  - Vibration
  - Joint position sense
- Cranial nerves
  - Visual acuity
  - Fields
  - Pupils
  - Eye movements   Nystagmus
  - Facial movements
  - Facial sensation
  - Hearing
  - Speech and swallowing

Summary/ Impression

Differential Diagnoses

Plan

History and examination crib sheet – with headings only

History
Presenting complaint

History of presenting complaint

Effect on daily life

Ideas

Concerns

Expectations

Drugs

Allergies

Past medical history

Family history

Social history

Systems enquiry
Cardiovascular system

Respiratory system

Gastrointestinal system

Genitourinary system

Central nervous system

Other

Summary

Differential diagnoses

Plan

INDEX

Printed in Great Britain
by Amazon